Text Book of

OPTICAL FIBER COMMUNICATION

For

Third Year Diploma – Semester VI

Diploma in Electronics and Communication Engineering

As Per New Syllabus of SBTE, Jharkhand

Mrs. Pratibha D. Kulkarni
M.Tech., (E & TC)
Sr. Lecturer, E & TC Department
Pimpri-Chinchwad Education Trust's
Pimpri-Chinchwad Polytechnic.
Nigdi, PUNE - 44.

Miss Sharvari D. Kulkarni
M.E. (E & TC)

NIRALI PRAKASHAN
ADVANCEMENT OF KNOWLEDGE

N4747

Optical Fiber Communication **ISBN 978-93-89944-05-1**

First Edition : February 2020

Published By :

NIRALI PRAKASHAN

Abhyudaya Pragati, 1312, Shivaji Nagar

Off J.M. Road, PUNE – 411005

Tel - (020) 25512336/37/39, Fax - (020) 25511379

Email : niralipune@pragationline.com

➢ DISTRIBUTION CENTRES

PUNE

Nirali Prakashan (For orders within Pune) : 119, Budhwar Peth, Jogeshwari Mandir Lane, Pune 411002, Maharashtra
Tel : (020) 2445 2044, Mobile : 9657703145
Email : niralilocal@pragationline.com

Nirali Prakashan (For orders outside Pune) : S. No. 28/27, Dhayari, Near Asian College Pune 411041
Tel : (020) 24690204 Fax : (020) 24690316; Mobile : 9657703143
Email : bookorder@pragationline.com

MUMBAI

Nirali Prakashan : 385, S.V.P. Road, Rasdhara Co-op. Hsg. Society Ltd.,
Girgaum, Mumbai 400004, Maharashtra; Mobile : 9320129587
Tel : (022) 2385 6339 / 2386 9976, Fax : (022) 2386 9976
Email : niralimumbai@pragationline.com

➢ DISTRIBUTION BRANCHES

JALGAON

Nirali Prakashan : 34, V. V. Golani Market, Navi Peth, Jalgaon 425001, Maharashtra,
Tel : (0257) 222 0395, Mob : 94234 91860; Email : niralijalgaon@pragationline.com

KOLHAPUR

Nirali Prakashan : New Mahadvar Road, Kedar Plaza, 1st Floor Opp. IDBI Bank, Kolhapur 416 012
Maharashtra. Mob : 9850046155; Email : niralikolhapur@pragationline.com

NAGPUR

Nirali Prakashan : Above Maratha Mandir, Shop No. 3, First Floor,
Rani Jhanshi Square, Sitabuldi, Nagpur 440012, Maharashtra
Tel : (0712) 254 7129; Email : niralinagpur@pragationline.com

DELHI

Nirali Prakashan : 4593/15, Basement, Agarwal Lane, Ansari Road, Daryaganj
Near Times of India Building, New Delhi 110002 Mob : 08505972553
Email : niralidelhi@pragationline.com

BENGALURU

Nirali Prakashan : Maitri Ground Floor, Jaya Apartments, No. 99, 6th Cross, 6th Main,
Malleswaram, Bengaluru 560003, Karnataka; Mob : 9449043034
Email: niralibangalore@pragationline.com

Other Branches : Hyderabad, Chennai

Dedication ...

Dedicated to my beloved Father

Shri. Krishnaji A. Deshpande (Principal)

Mother

Late Sau. Sunita K. Deshpande

– Mrs. Pratibha D. Kulkarni

Dedicated to my beloved Father

Shri. Dipak M. Kulkarni

Mother

Sr. Lecturer Sau. Pratibha D. Kulkarni

– Miss Sharvari D. Kulkarni

Preface ...

It gives me great pleasure and immense satisfaction in presenting this text book of '**Optical Fiber Communication** VIth Semester of Third Year Diploma Course in Electronics and Telecommunication Engineering students pertaining to SBTE's, Jharkhand.

This subject is presented in a simple language considering the requirements of all the students regarding the changing trends of examination. Utmost care has been taken to check the mistakes and misprints, yet it is very difficult to claim perfection.

I take this opportunity to thank **Hon. Secretary Shri. V. S. Kalbhor, Hon. A. O. Shri. Vispute** and **Hon. Principal Mrs. V. S. Byakod, Vice-Principal Hon. Shri. B. V. Mane and Shri. Gorakh Bhalekar** for their moral support and encouragement.

I am also thankful to the publisher **Shri. Dineshbhai Furia** and **Shri. Jignesh Furia**. I am also thankful to Mr. Malik Shaikh, Mrs. Anagha Medhekar (Co-ordination and Proof Reading), Mrs. Anjali Muley (Figure Drawing) and staff of Nirali Prakashan for bringing out this book.

A very special thanks my father **Shri. Dipak Kulkarni**, Brother **Chi. Shriprasad** and Husband **Shri. Mangesh Gijare** for their support and co-operation.

Any undetected and unintentional error, omissions, suggestions etc. from students and teachers for improvement brought to our notice in good spirit are most welcome.

Pune.

Mrs. Pratibha D. Kulkarni
Miss Sharvari D. Kulkarni

Syllabus ...

Unit-I Introduction **(6 Hrs.)**

Need of Fiber Optic Communications, Evolution of Light wave Systems, Basic Concepts; Analog and Digital Signals, Channel Multiplexing, Modulation Formats, Optical Communication Systems, Light wave System Components; Optical Fibers as a Communication Channel, Optical Transmitters, Optical Receivers.

Unit-II Optical Fibers Geometrical **(10 Hrs.)**

Optics Description; Step-Index Fibers, Graded Index Fibers, Wave Propagation; Maxwell's Equations, Fiber Modes, Single-Mode-Fibers, Dispersion in Single-Mode Fibers; Group Velocity Dispersion, Material Dispersion, Wave guide Dispersion, Higher-order Dispersion, Polarization-Mode Dispersion, Dispersion-Induced Limitations; Basic Propagation Equation, Chirped Gaussian Pulses, Limitations on the Bit Rate, Fiber Bandwidth, Fiber Losses; Attenuation Coefficient, Material Absorption, Rayleigh Scattering, wave guide Imperfections, Non-linear Optical effects; Stimulated Light Scattering, Non-linear Phase Modulation, Four Wave Mixing, Fiber Manufacturing; Design Issues, Fabrication Methods, Cables and Connectors

Unit-III Optical Transmitters Basic Concepts; **(8 Hrs.)**

Emission and Absorption Rates, p-n Junctions, Non-radiative Recombination, Semi conductor Materials, Light Emitting Diodes; Power-current Characteristics, LED spectrum, Modulation Response, LED Structures, Semi Conductor Lasers; DFB Lasers, Coupled Cavity semiconductor Lasers, Tunable Semiconductor Lasers, Vertical Cavity Semiconductor Lasers, Laser Characteristics, Small and Large Signal Modulation, Spectral Line width, Source Fiber Coupling.

Unit-IV Optical Receivers **(8 Hrs.)**

Basic concepts, p-n Photo Diodes, p-i-n Photo Diodes, Avalanche Photo Diode, MSM Photo detector, Receiver Design, Receiver Noise; Noise mechanism, Receiver sensitivity; Bit error rate, Minimum Receiver Power, Sensitivity Degradation, Receiver Performance.

Unit-V Light Wave Systems **(5 Hrs.)**

System Architecture, Loss limited Light wave systems, Dispersion limited Light wave systems, Power Budget, Long Haul systems, Sources of Power Penalty; Model Noise, Dispersive Pulse Broadening, Mode Partition Noise, Frequency Chirping, Reflection Feedback Noise

Unit-VI Multi channel Systems **(5 Hrs.)**

WDM Light wave systems, Optical TDM Systems, Subscriber Multiplexing, Code Division Multiplexing.

***

Contents ...

***

Chapter 1

INTRODUCTION

1.1 INTRODUCTION

- The communication medium in most electronic communication systems is either a wire conductor cable or free space.
- One of the main limitations of communication systems is their restricted information carrying capabilities.
- As we know, the information handling ability is directly proportional to the bandwidth of the communications channel.
  For example, in telephonic systems, the bandwidth is limited by the characteristics of the cable used to carry the signals.
- As the demand for telephones has increased, better cables and wiring systems have been developed.
- Further, multiplexing techniques have been developed to transmit multiple telephone conversations over a single cable.
- These techniques have the same effect as the number of cables or channels of communications were greatly multiplied.
- But now-a-days, a new medium is growing in popularity that is fiber optic cable.
- A fiber optic cable is a glass pipe used to carry a light beam from one place to another.
- Light is an electromagnetic signal like a radio wave.
- It can be modulated by information and sent over the fiber optic cable.
- Because the frequency of light is extremely high, it can accommodate very wide bandwidths of information and extremely high data rates can be achieved with excellent reliability.
- Because of rapidly increasing demands for telephone communication throughout the world, multiconductor copper cables have become not only very expensive but also insufficient way to meet these information requirements.
- The frequency limitations in the copper conductor system approximately 1 MHz.
- The optical fiber with its low weight and high frequency characteristics approximately 40 GHz and its imperviousness to interference from electromagnetic radiation, have become the choice for all heavy demand long-line telephone communication systems.

1.2 OPTICAL SPECTRUM

- Light is a kind of electromagnetic radiation.
- Another familiar form of electromagnetic radiation is a radio wave.
- Any electromagnetic wave is made-up of both electric and magnetic fields that travel through space from one place to another place.
- The basic important characteristics of electromagnetic radiation is its frequency or wavelength.
- To put light waves into perspective with the lower radio and other frequencies shown in Fig. 1.1.

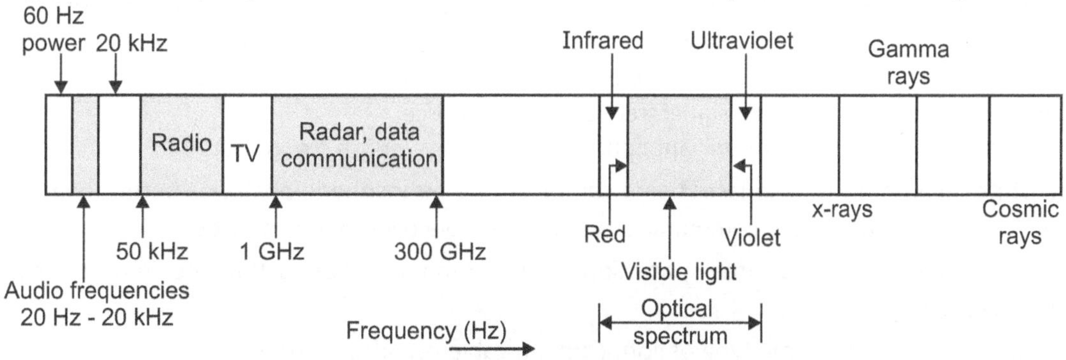

Fig. 1.1: Electromagnetic Frequency Spectrum

- At left end of the spectrum are very low frequency electrical signals such as 60 Hz ac power and audio frequencies in the range of 20 Hz to 20 kHz.
- These are low frequencies that have long wavelengths.
- Higher up the scale to the right side is a wide range of radio frequencies.
- Radio frequency range is from 10 kHz to 300 GHz (approx.).
- Microwaves extend from 1 GHz to 300 GHz.
- Even further up the scale we find visible light.
- **This visible light range between infrared and ultraviolet rays is called optical spectrum, of the range 3 × 10" to 3 × 10^6 Hz.**
- This includes both infrared and ultraviolet as well as visible parts of the spectrum.
- The visible spectrum ranges from 4.3×10^{14} to 7.5×10^{14} Hz, but we rarely refer to the frequency of light.
- Instead, we state light in terms of wavelength.
  (Where, wavelength is a distance measured in meters between peaks of a wave).
- Light waves are very short and are usually expressed in nanometers or micrometers or microns.
- Visible light is in the range of 400 to 700 nanometers (nm) or 0.4 to 0.7 μm (micrometers), depending on the colour of light.
- Short wavelength light is violet (400 nm) and real (700 nm) is long-wavelength light.
- Right below visible light is a region known as infrared.
  (Infrared rays cannot be seen).
- However, they have the properties of light and therefore, can be manipulated in the same way with mirrors, lenses and other devices.
- The source of most infrared is heat.
- Right above the visible light is the ultraviolet range.
- Infrared and ultraviolet are included in optical spectrum.

1.3 HISTORY OF FIBER OPTIC

- One way to expand communication capability further is to use light as the transmission medium.
- Instead of using an electrical signal travelling over cable or electromagnetic waves travelling through space, the information is put on a light beam and transmitted through space or through a special cable.
- The use of visible optical carrier waves or light for communication has been for many years.
- Simple systems such as signal fires, reflecting mirrors and more recently, signaling lamps have provided successful if limited information transfer.
- Moreover, as early as 1880 **Alexander Graham Bell,** the inventor of the telephone, demonstrated that information could be transmitted by light.
- A special transmitter made-up of a microphone attached to a mirror was used to transmit voice over a distance of several hundred feet using a light beam from the sun.
- However, although some investigation of optical communication continued in the early part of the twentieth century and its use was limited to mobile, low capacity communication links.
- Communication by light beam in free space is impractical over very long distances because of great attenuation of the light due to atmospheric effects like fog, haze, smog, rain, snow and other conditions absorb, reflect and refract the light, greatly attenuating it and thereby limiting the transmission distance.
- Artificial light beams used to carry information are also virtually obliterated during day light hours by the sun.
- Light beam communications was made more practical in the early 1960's with the invention of the **laser.**
- The laser is special height-intensity, signal frequency light source.
- It produces a very narrow beam of brilliant light of specific wavelength (i.e. colours).
- Because of its great intensity, the laser beam can penetrate atmospheric obstacles better than other types of light, thereby making light beam communication more reliable over long distances.
- The primary problem with such free-space light beam communication is that the transmitter and receiver must be perfectly aligned.
- Instead of using free space, some type of light carrying cable can also be used.
- For centuries, it has been known that light is easily transmitted through various types of transparent media such as glass and water.

- By the mid-1950's, glass fibers were developed that permitted long light-carrying cables to be constructed.
- Over the years these glass fibers have been perfected.
- Further, low-cost plastic fiber cable was also developed.
- Today, these fiber optic cables have been highly refined.
- Cables may miles long can be constructed and interconnected for the purpose of transmitting information on a light beam over very long distances.
- These highly refined fiber optic cables, a new transmission medium is now available.
- Its great advantage is that the light beam have an incredible information capacity.
- Whereas hundreds of telephone conversations may be transmitted simultaneously at microwave frequencies thousands of signals can be carried on a light beam through a fiber optic cable.
- In parallel with the development of the fiber waveguide attention was also focused on the other optical components.
- Since optical frequencies are accompanied by extremely small wavelengths the development of all these optical components required a new technology.
- Thus, semiconductor optical sources (Laser and LED) and detectors (photodiodes and phototransistors) compatible in size with optical fibers were designed and fabricated to enable successful implementation of the optical fiber system.
- Initially, the semiconductor lasers exhibited very short life times of at best a few hours, but significant advances in the device structure enabled life times greater than 1000 hr and 700 hr to be obtained by 1973 and 1977 respectively.
- These devices was originally fabricated from alloys of (gallium arsenides (AlGaAs) which emitted in the near infrared between 0.8 and 0.9 μm.
- To obtain both low loss and low dispersion at the same operating wavelength, new advanced single mode fiber structures have been realized, namely, dispersion shifted and dispersion flattened fibers.
- Hence, developments in fiber technology have continued rapidly over recent years.
- High performance, reliable optical fiber communication systems are now widely deployed both within telecommunications networks and many others more localized communication application areas.

1.4 BASIC CONCEPTS OF ANALOG DIGITAL SIGNALS

1. **Analog signal**

   Analog signal is the signal, in which voltage or current varies continuously with time.

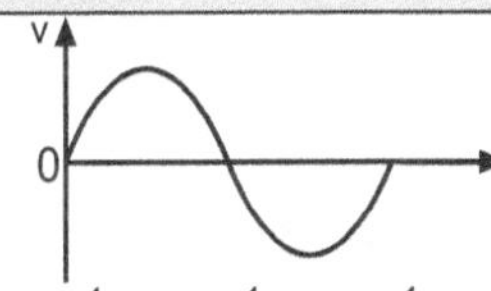

2. **Digital signal**

   Digital signal is the discontinuous signal, which has only two levels, high and low or one and zero.

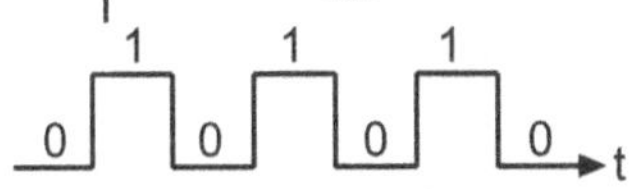

1.4.1 Channel Multiplexing

- Multiplexing divides the physical line or a medium into logical segments called channels.
- In multiplexing, different channels carry data simultaneously over same physical medium.
- Hardware equipment called multiplexer combine the inputs from different sources and load them on different channels of a medium.
- The combined data transverses over the medium simultaneously.
- At the receiving end, a demultiplexer separates the signals and sends them to their different destinations.
- Here, we have to study multiplexing techniques like TDM, FDM, WDM

1.4.1.1 Multiplexing Techniques

- Whenever the bandwidth of a medium linking two devices if greater than the bandwidth needs of the devices, then the link can be shared.
- *Multiplexing is the set of techniques that allows the simultaneous transmission of multiple signals across a single data link.*
- As data and telecommunication usage increases, so does the traffic also.
- We can install higher bandwidth links and use each to carry multiple signals.

- If the bandwidth of a link is greater than the bandwidth needs of the devices connected to it, the bandwidth is wasted.
- An efficient system maximizes the utilization of all resources, bandwidth is one of all the most precious resources we have in the data communication.
- In multiplexed system, n lines share the bandwidth of one link.
- Fig. 1.2 shows the basic format of a multiplexed system.

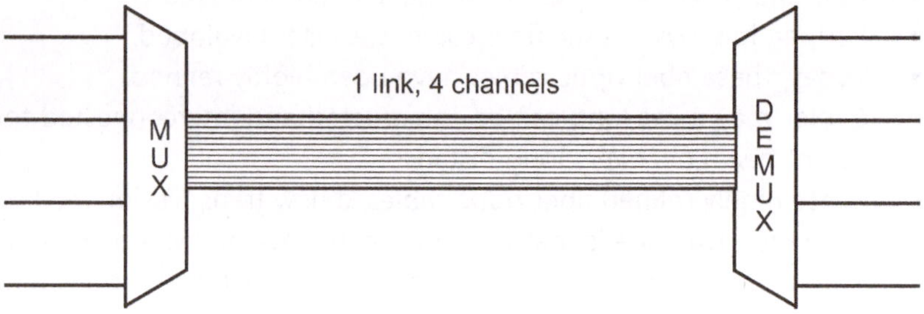

Fig. 1.2 : Dividing a Link into Channels

- **Multiplexer** combines the four lines into one (many-to-one).
- At the receiving end, **Demultiplexer** separates the stream into four lines (one-to-many).

Need of Multiplexing :

- When many signals (channels) to be transmitted, then from transmitter side that sends simultaneously i.e. multiplexer coverts many into one so that at the receiving end also all inputs we get simultaneously sending many signals separately is expensive and requires more wires to send.
  For example, In cable TV, distributor sends many channels through single wire.

Types of Multiplex are :

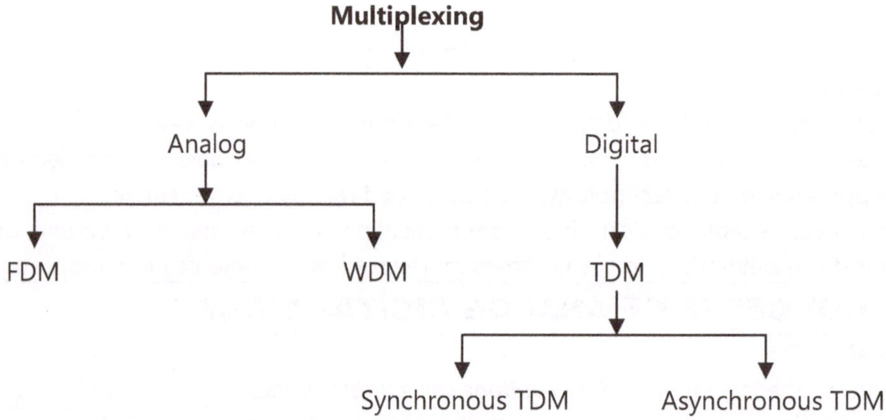

1.4.2 Frequency-Division Multiplexing (FDM)

- It is an analog technique.
- *Frequency division multiplexing means total range of frequency is divided into number of frequency slots. Each slot of frequency is allotted to each channel.*

Principle :

Various channels of different frequencies combined, transmitted through single wire and separated at receiver with the help of demultiplexer.

- FDM can be applied when the bandwidth of a link is greater than the combined bandwidths of the signals to be transmitted.
- These modulated signals are then combined into a single composite signal that can be transported by the link.
- Carrier frequencies are separated by sufficient bandwidth to accommodate the modulated signal.
- These bandwidth ranges are the channels through which the various signals travels.
- Channels must be separated by guard bands to prevent signals from overlapping.
- Fig. 1.3 shows the conceptual view of FDM.

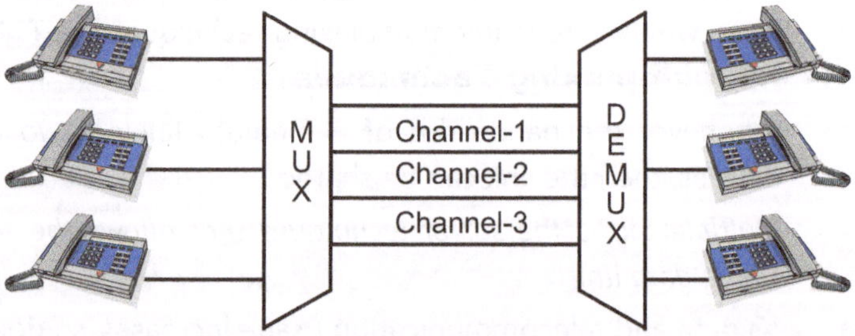

Fig. 1.3 : FDM

From Fig. 1.3, the transmission path is divided into three parts, each representing a channel to carry one transmission.

Example, three narrow streets merge to form a three lane highway. Each of three streets corresponds to a lane of highway. Each car merging into the highway from one of the streets still has own lane and can travel without interfering with other cars.

1.4.2.1 FDM Transmitter (Multiplexing Process)

- As shown in Fig. 1.4, FDM is an analog process.
- Each signal is of similar frequency range.
- Each signal modulates a separate carrier and a modulator contains the sidebands of the corresponding signal.
- All the outputs from modulator added together in adder.
- Adder is a linear mixer which gives algebraic sum of all modulated signals.
- The resulting modulated signals are then combined into a signal that is sent over media link that has enough bandwidth to accommodate it.

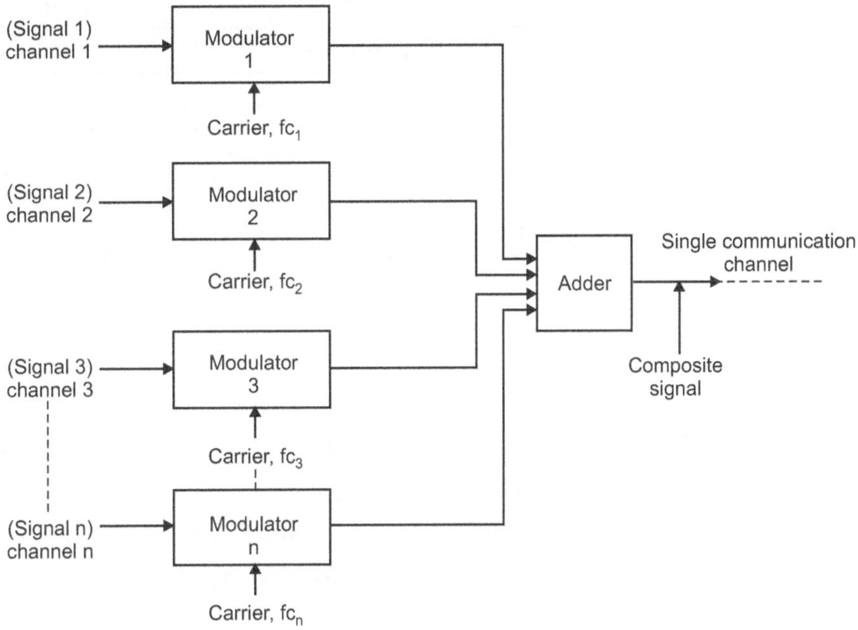

Fig. 1.4 : FDM Transmitter

1.4.2.2 FDM Receiver

- Demultiplexing process done at receiver.
- Demultiplexing means one into many. The composite signal is applied to different bandpass filters which separates individual channel.
- Each BPF has a center frequency corresponding to one of the carrier frequency.
- Each BPF pass only its channel and rejects all other channels.
- Each demodulator removes the corresponding carrier frequency and recovers the original signal back.
- The block diagram of FDM receiver is shown in Fig. 1.5.

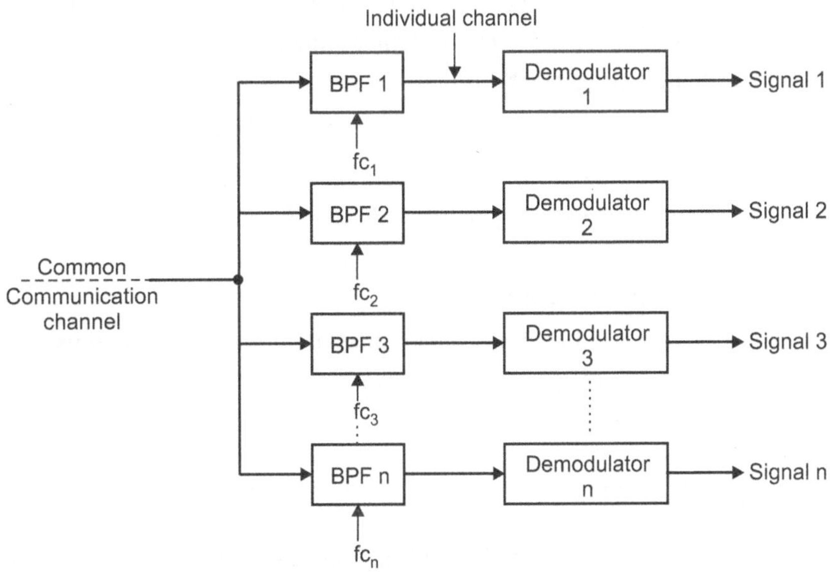

Fig. 1.5 : FDM Receiver

Example 1.1 :

Assume that a voice channel occupies a bandwidth of 4 kHz. We need to combine three voice channels into a link with a bandwidth of 12 kHz, from 20 to 32 kHz. Show the configuration using the frequency domain. Assume that there are no guard bands.

Solution :

We modulate each of three voice channels to a different bandwidth, as shown in Fig. 1.6.

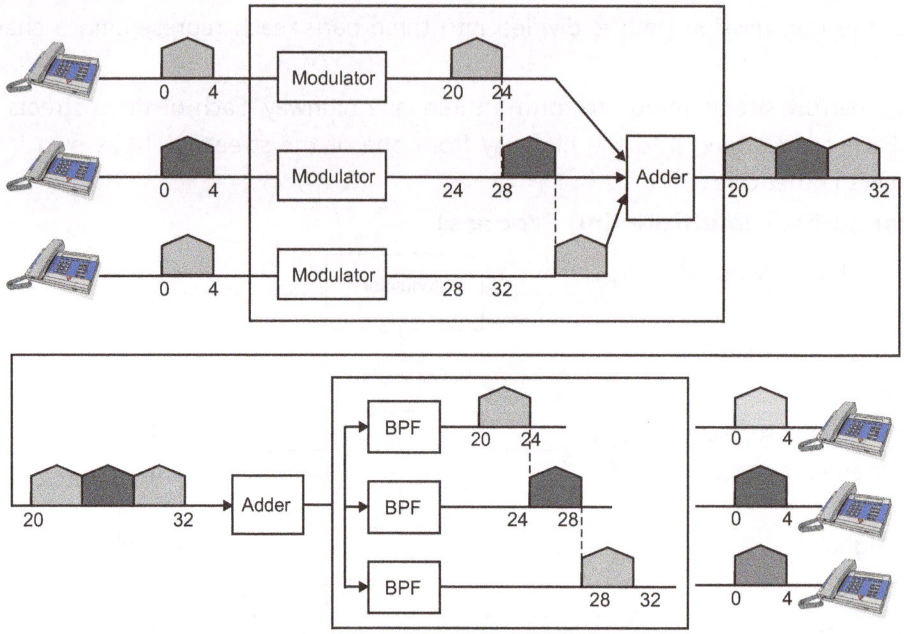

Fig. 1.6

- We use 20 to 24 kHz bandwidth for 1st, 24 to 28 kHz for 2nd and 28 to 32 kHz for 3rd channel.
- Then we combine these shown in Fig. 5.5.
- At the receiver, each channel receives the entire signal using a filter to separate out its own signal.

1.4.2.3 Advantages and Disadvantages and Applications of FDM

- **Advantages of FDM :**
  1. FDM is simpler and easy demodulation.
  2. Used for analog signals.
  3. Does not need synchronization between its transmitter and receiver.
  4. Due to slow narrow band fading only one channel gets affected.
- **Disadvantage of FDM :**
  1. Used only when a few low speed channels are desired.
  2. The communication channel must have large bandwidth.
  3. Intermodulation distortion takes place.
  4. FDM suffers problem of cross-talk.
- **Applications of FDM :**
  1. FDM is used to public telephones and in cable TV systems.
  2. AM and FM radio broadcasting.
  3. TV broad-casting.
  4. First generation of cellular phones.

Example 1.2 :

Five channels, each with a 100 kHz bandwidth, are to be multiplexed together. What is minimum bandwidth of the link if there is a need for a guard band of 10 kHz between the channels to prevent interference ?

Solution :

For five channels, 4 guard bands are required.

$$\therefore \quad \text{Required bandwidth} = 5 \times 100 + 4 \times 10$$

$$= 540 \text{ kHz}$$

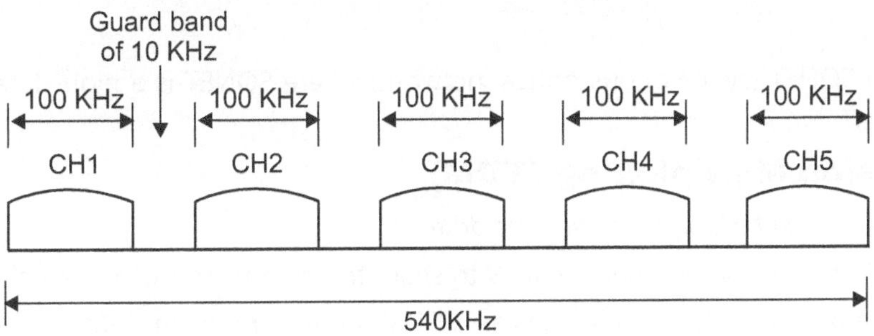

Fig. 1.7

1.4.3 Wave-Division Multiplexing (WDM)

- *WDM is an analog multiplexing technique to combine optical signals.*

Principle :

Very narrow bands of light from different sources are combined to make a wider band of light and at the receiver, the signals are separated by the demultiplexer.

- WDM is designed to use the high data rate capability of fiber optic cable.
- The optical fiber data rate is higher than the data rate of metallic transmission cable.
- Using a fiber optic cable for one single line wastes the available bandwidth.
- Multiplexing allows us to connect several lines into one.
- WDM is conceptually the same as FDM, except that the multiplexing and demultiplexing involve the optical signals transmitted through fiber optic cable.
- Fig. 1.8 shows the conceptual view multiplexer and demultiplexer.

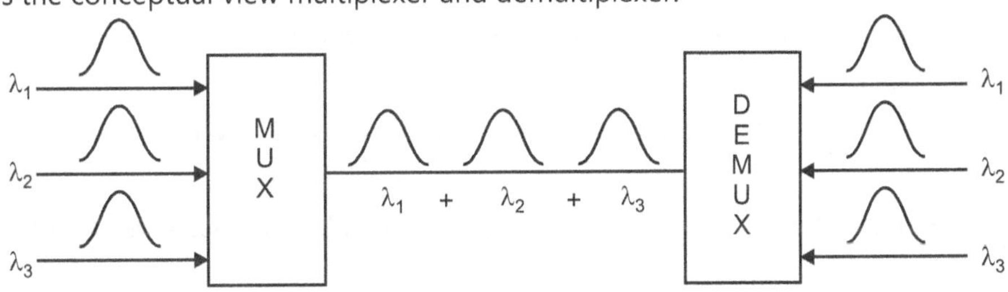

Fig. 1.8 : WDM System

- Very narrow bands of light of differential wavelengths are combined to make a wide band of light.
- All wavelength travels through signal cable.
- At the receiver, the signals are separated by the demultiplexer.
- Combining and splitting of light sources are easily handled by a prism.
- We studied in Physics, that a prism bends a beam of light based on the angle of incidence and the frequency.
- Using this technique, a multiplexer can be made to combine several input beams of light, each containing narrow band of frequencies into one output beam of wider band of frequencies.
- A demultiplexer does the reverse process.

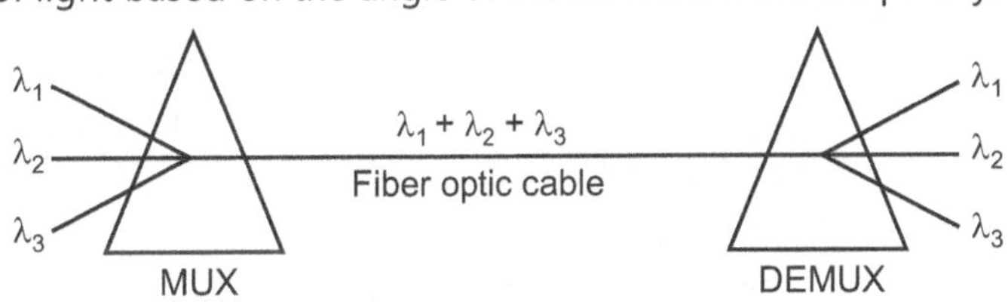

Fig. 1.9 : Prisms with WDM

- **Advantages of WDM :**
  1. In low bandwidth large number of channels accommodated.
  2. Time sharing not required.
  3. High rate.
  4. High speed.
- **Disadvantage of WDM :**

  Only possible with fiber optic cable which is expensive.

Applications of WDM :

- WDM used in SONET (synchronous optical network) where SONET is a high-data-rate network for fiber optic networks.

1.4.4 Time-Division Multiplexing (TDM)

- *TDM is digital multiplexing technique to combine data.*
- TDM is a process that allows several connections to share the high bandwidth of a link.
- Instead of sharing a portion of the bandwidths as in FDM, time is shared in TDM.

Principle :

Various channels having different time slots combined in a single wire and at the receiver they are separated with the help of demultiplexer.

- Fig. 1.10 shows conceptual view of TDM.

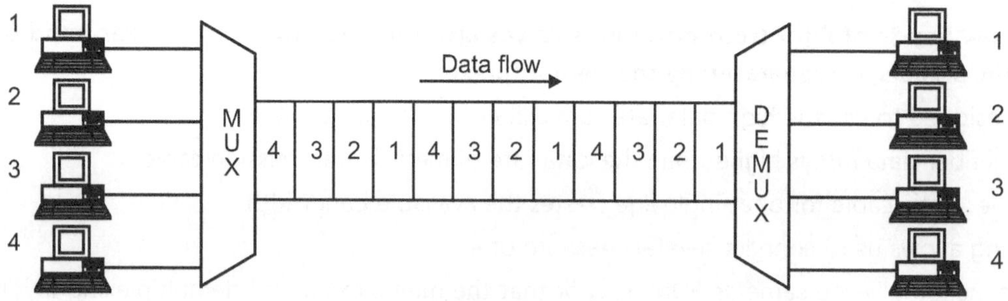

Fig. 1.10 : TDM

- Same link is used as in FDM, but the link is shown sectioned by time rather than by frequency.
- In Fig. 1.10 portions of signals 1, 2, 3 and 4 occupy the link sequentially.
- We divide the transmission time into a number of time slices.
- We then allocate each time slice to different source nodes, each of which wants to send data.

Time Slots and Frames :

- The data flow of each connection is divided into units and the link combines one unit of each connection to make a frame.
- The size of the unit can be 1-bit or several bits. Fig. 1.10 shows an example.
- In TDM, the data rate of the link that carries data from n connections must be n times the data rate of a connection to guarantee the flow of the data.
- Thus, the duration of a unit in a connection in n times the duration of a time slot in a frame.

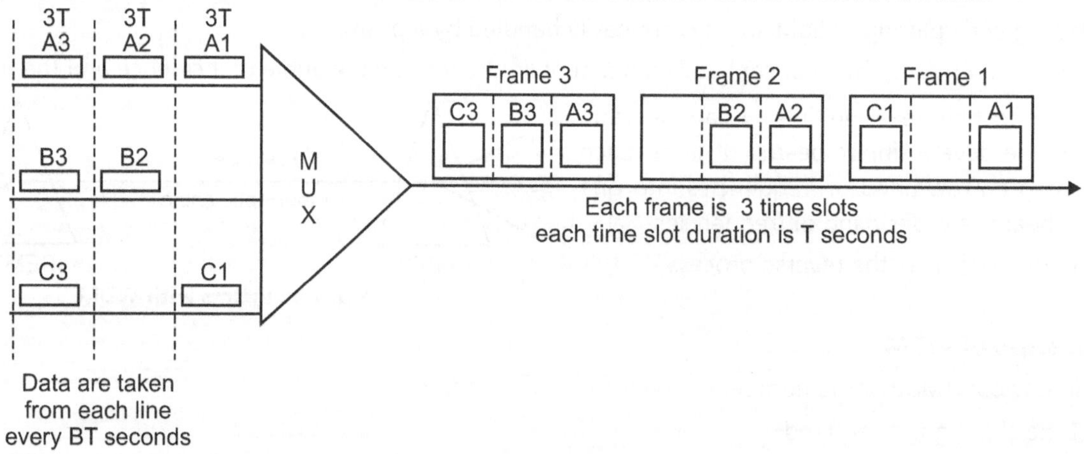

Fig. 1.11 : TDM

- As shown in Fig. 1.11, the data rate of the link is 3 times the data rate of connection such that the duration of a unit on a connection is 3 times that of time slot.
- In Fig. 1.11, the data prior to multiplexing as 3 times the size of the data after multiplexing.

- This is just to convey the idea that each unit is 3 times longer duration before multiplexing than after.
- Time slots are grouped into frames.
- Here frame consists of one complete cycle of time slots with one slot decided to each sending device.
- In this system input lines are n, each frame has n slots, with each slot allocated to carrying data from a specific input line.

In such case, each node may want different amounts of data to be transmitted, because the time slice allocated to each source would be different. This requires a large number of buffer at each node to hold data when some other source node is transmitting. This method can be unsuitable to a few source, nodes at it may involve long waits if one of the source node **graphs** the medium for longer time.

For this reason or to overcome this problem the time span is divided into equal time slices.

So the TDM can be implemented in two ways.

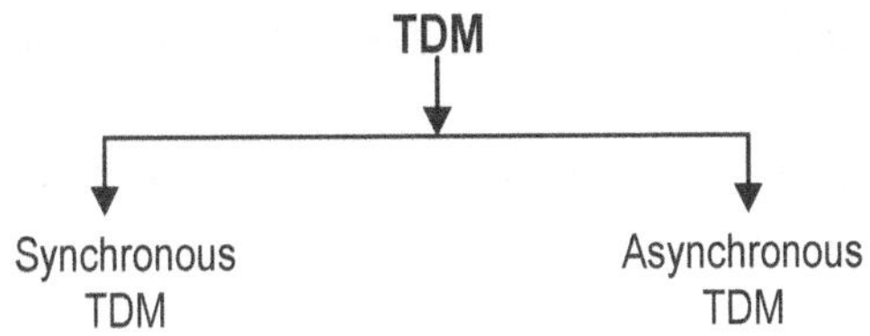

1.4.4.1 Synchronous TDM

The synchronous TDM also referred to as TDM.

The meaning of synchronous is different in other areas of telecommunication. But here it means that the multiplexer allocates exactly the same time slots to each device at all times, whether or not a device has anything to transmit.

For example, time slot A, is assigned to device A alone and not to used for any other device.

Fig. 1.12 shows synchronous TDM.

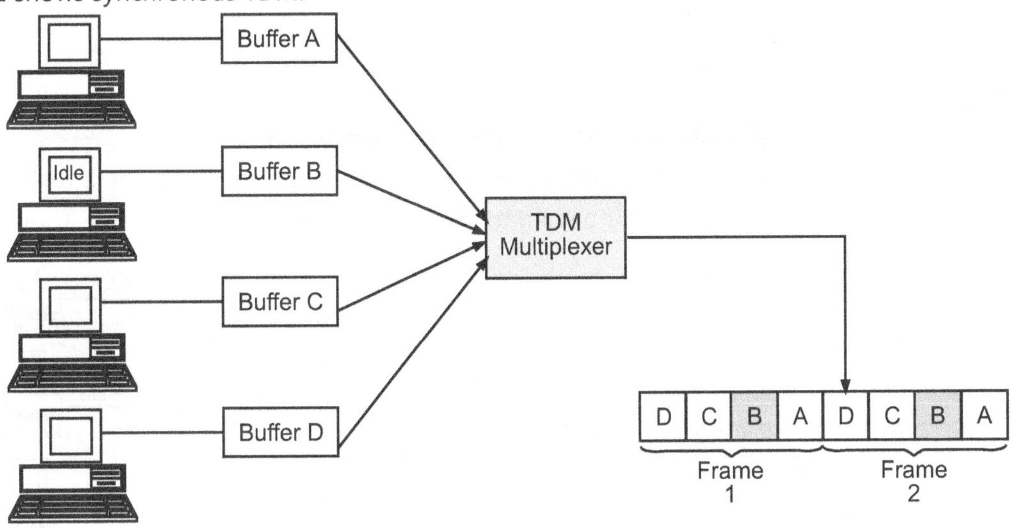

Fig. 1.12 : Synchronous Time Division Multiplexing

- A small buffer memory is associated with each source node.
- At any time, not all the nodes may want to send some data.
- But the timing device in the multiplexer allocates some time for each node to transmit the data from its buffer as shown in figure.
- These time slots are grouped into frames. A frame consists of one complete cycle of time slots.

  For example, for n inputs, each frame has at least n slot, with each slot allocated to carrying data from a specific input line.
- Each device has one time slot per frame.
- After transmitting the data, the cycle repeats i.e. A-B-C-D-A-B-C-D, etc.

- Then next turn comes, if a node wants to transmit any data.
- If there is no data to be transmitted, the buffer will be empty, but still the turn of the node will come shown in Fig. 1.12 (as node B).
- Here no addressing scheme to identify the source of the data but the position of the data in the composite message determine the source address.

- **Interleaving :**
  - Synchronous TDM can be composed to a very fast rotating switch.
  - As the switch opens in front of a device, that the device has the opportunity to send a specified amount of (bits) of data onto the path.
  - The switch moves from device to device at a constant rate and in a fixed order. This process is called **interleaving**.

Interleaving can be done by bit, byte or any other data unit. The care must be taken that the interleaved units will always be of the same size.

Fig. 1.13 shows interleaving and frame building. Each device is sending different message. The multiplexer interleaves the different messages and forms them into frames before putting them onto the link.

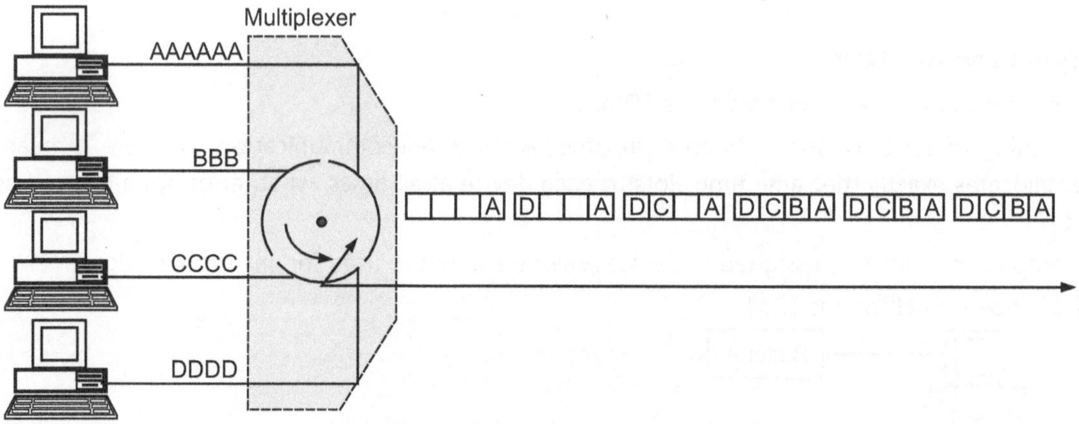

Fig. 1.13 : Synchronous TDM, Multiplexing Process

At the receiver, the demultiplexer decomposes each frame by extracting each character in turn.

As a character is removed from a frame, it is passed to the appropriate receiving device as shown in Fig. 1.14.

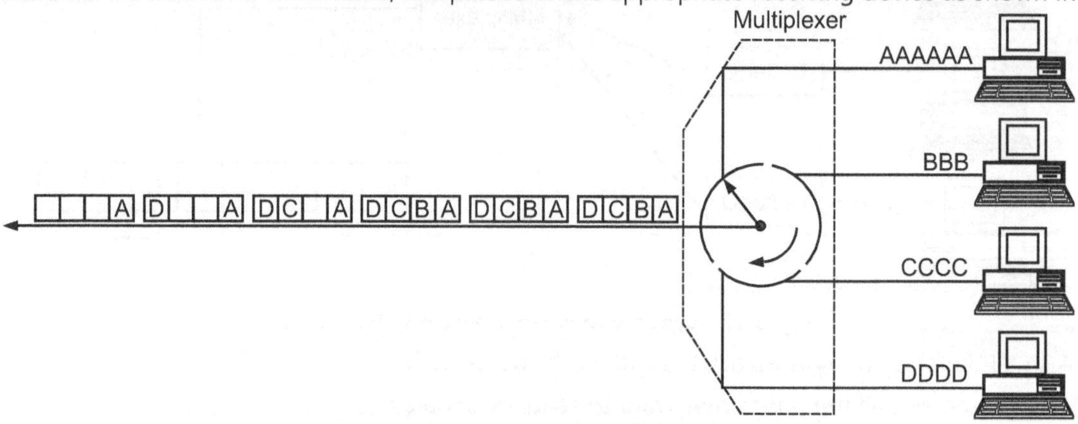

Fig. 1.14 : Synchronous TDM, Demultiplexing Process

1.4.4.2 Asynchronous TDM

Here asynchronous means flexible or not fixed.

Asynchronous TDM is also known as **Statistical TDM** (STDM).

- This technique is more intelligent.
- It monitors which computer or terminal sending data more frequently and in greater quantity and allocates the time slices to those nodes.

- Inactive terminals/computers get the time slice less, while completely idle computer/terminal may not get any time slice at all.
- Only statistical data is kept about the activity of various source nodes and hence the name.

Fig. 1.15 shows the asynchronous TDM where the node B does not send any data at all while node C is more active, hence is **polled** more frequently.

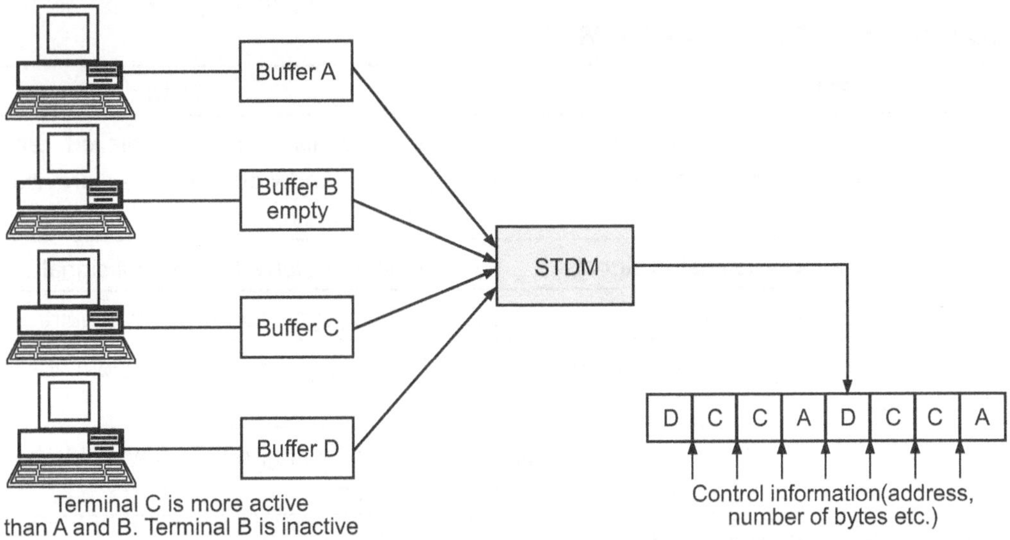

Fig. 1.15 : Statistical Time Division Multiplexing (STDM)

1.4.4.3 Synchronizing

- Implementation of TDM is not easy as that of FDM.
- TDM needs synchronization between multiplexer and demultiplexer.
- If synchronization is not there between multiplexer and demultiplexer, a bit belonging to one channel may be received by the wrong channel.
- For this reason, one or more synchronization bits are usually added to the beginning of each frame.
- These bits are called **framing bits,** follow a pattern, frame to frame, that allows the demultiplexer to synchronize with the incoming stream so that it can separate time slot accurately.
- In most of the cases, this synchronization information consists of 1-bit per frame, changing between 0 and 1 as shown in Fig. 1.16.

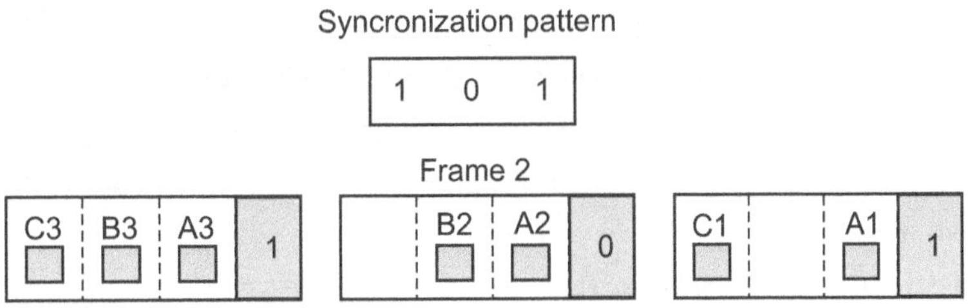

Fig. 1.16 : Framing Bits

1.4.4.4 Advantages, Disadvantages and Applications of FDM and TDM

- **Advantages of TDM :**
  1. TDM systems are more flexible than FDM.
  2. Digital systems so often noise.
  3. No intermodulation distortion.
  4. Problem of cross-talk is not severe.
- **Disadvantage of TDM :**
  1. Complex to implement.
  2. Synchronization required.

- **Application of TDM :**
  1. In ISDN telephone lines.
  2. In wireline telephone systems.
  3. Some cellular telephone system.
  4. In circuit switched networks such as Public Switched Network (PSTN).

1.4.5 Comparison of FDM and TDM

FDM	TDM
1. The signals to be multiplexed are added in time domain. But they occupy different slots in the frequency domain.	1. The signals to be multiplexed can occupy the entire bandwidth but they are isolated in the time domain.
2. Mostly FDM is preferred for the analog signals.	2. TDM is preferred for digital signals.
3. In FDM, synchronization is not required.	3. In TDM, synchronization is required.
4. FDM suffers from cross-talk like problems due to imperfect band pass filter.	4. In TDM cross-talk like problem not occurs.
5. Due to wideband fading in the transmission medium, all the FDM channels are affected.	5. Due to fading, only a few TDM channels will be affected.

1.5 MODULATION FORMATES

- Modulation of binary data or **digital to analog modulation** is the process of changing one of the characteristics of an analog signal based on the information in a digital signal (0's and 1's).
- When we transmit data from one computer to another through a telephone line, the data must be converted into analog signal because the original data from computer is digital but telephone wires carry analog signals.
- The digital data must be modulated on an analog signal that corresponds to binary 1 and binary 0.
- Fig. 1.17 shows the relationship between the digital information, i.e. digital to analog modulation.

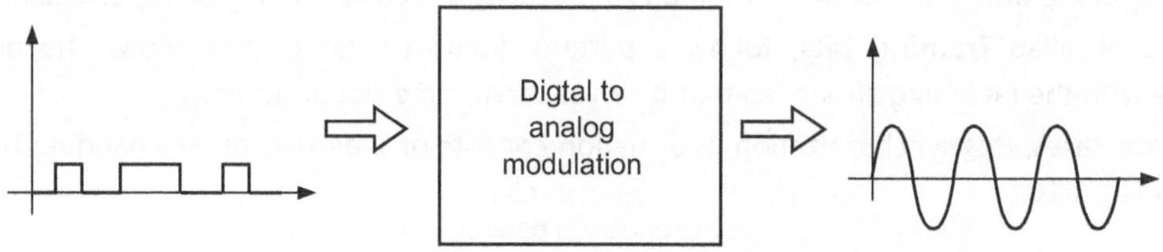

Fig. 1.17 (a) : Digital to Analog Modulation

- We know that sine wave has three variable parameters amplitude, frequency and phase.
- Corresponding to binary 1's and 0's one of the parameter of sine wave varies, this variation is in only two values so is known as shift keying.
- **Types of Digital-to-Analog Modulation :**

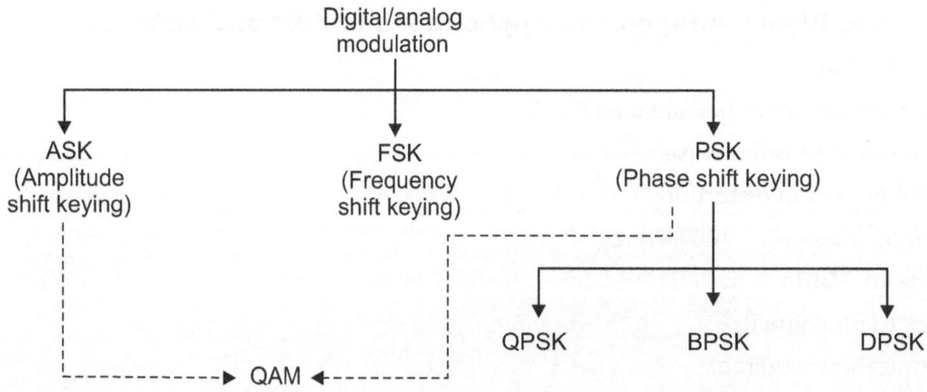

- Forth type that combines changes in both amplitude and phase called **Quadrature Amplitude Modulation (QAM).**
- QAM is the most efficient of these options and is the mechanism **used in all modern modems.**
  For example, modem is used for connecting a computer to telephone line.
- The telephone lines are designed to carry analog signals and their bandwidth is less, so they cannot be used for digital data transmission.
- A modem converts the digital data from computer to analog signals and puts it on the telephone lines.

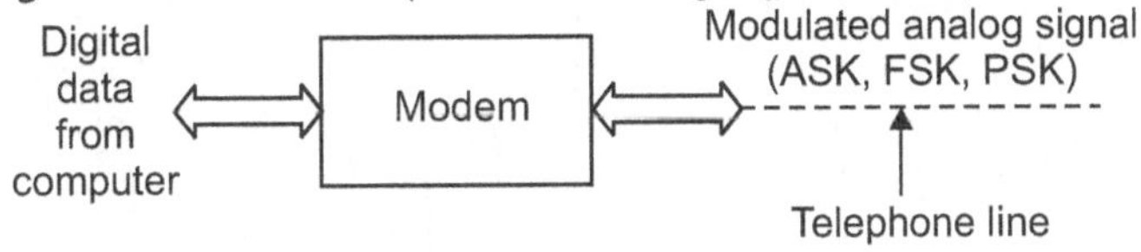

Fig. 1.17 (b) : Modem

- Modem means modulation and demodulation.
- The modulation techniques used for modem are – ASK, FSK and PSK.

Digital Modulation Schemes

Binary Schemes
- ASK
- FSK
- PSK

M-ary Schemes
- M-ary FSK
- M-ary PSK
- MSK
- QASK or QAM

- M-ary schemes need less bandwidth as compared to binary schemes, but the error performance of M-ary schemes is poor compared to binary schemes.

1.6 AMPLITUDE SHIFT KEYING (ASK)

- In **amplitude shift keying,** the amplitude of carrier is varied to represent binary 1 and 0. Both frequency and phase remains constant.
- i.e. in ASK amplitude shifts only in two levels high and low so known as amplitude shift keying.
- Fig. 1.18 (a) shows the block diagram of ASK transmitter.

1.6.1 ASK Transmitter

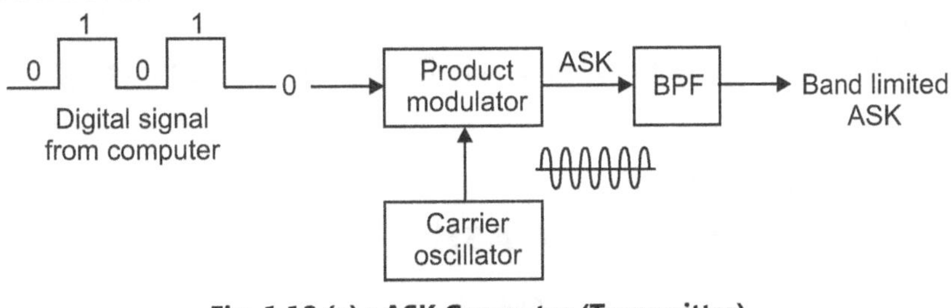

Fig. 1.18 (a) : ASK Generator (Transmitter)

- **Carrier oscillator :**
  Generates carrier i.e. sinewave with frequency f_c.
- **Digital signal :**
  Signal from computer acts as a modulating or information signal.
- **Product modulator :**
  It is a multiplier which multiplies the modulating signal and carrier.
- Due to multiplication, the ASK output will be present only when a binary '1' is to be transmitted.

- **BPF :**
  Band pass filter allows only wanted frequency.
- The ASK output corresponding to binary '0' is zero shown in Fig. 3.2 (b).

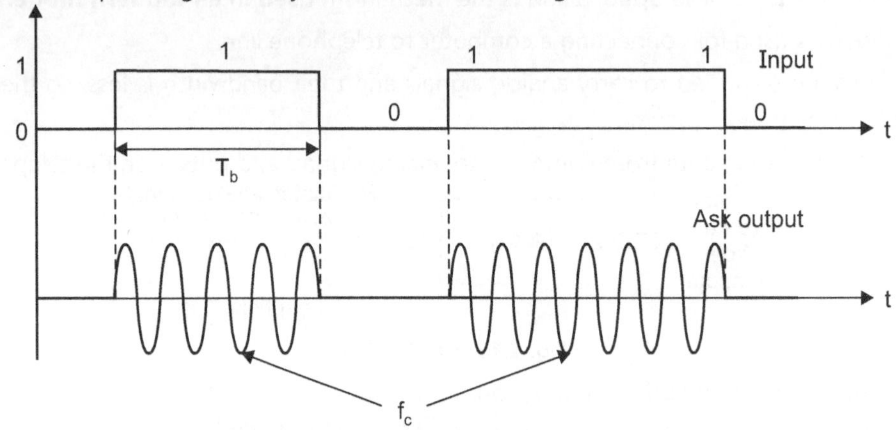

Fig. 1.18 (b) : ASK Input Output Waveforms

- The peak amplitude of the signal during each bit duration is constant and it's value depends on the bit (0 or 1).
- A popular ASK technique is called ON/OFF keying (00 K).

Bandwidth for ASK :

- The bandwidth of ASK signal is depends on bit rate R.

  where, bit rate $R = \frac{1}{T_b}$ shown in Fig. 1.18 (b).
- For bit rate 'R' bits/sec. bandwidth of ASK is

  $\boxed{B = 2R \text{ Hz}}$
- The frequency spectrum of ASK is shown in Fig. 1.18 (c).

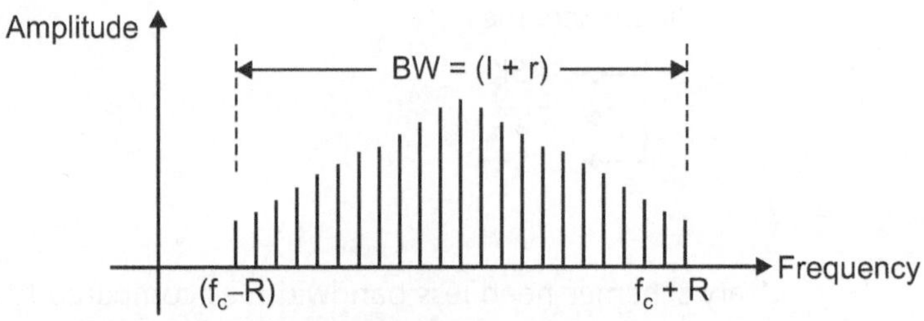

Fig. 1.18 (c) : Frequency Spectrum of ASK

- The transmission bandwidth of ASK signal can be restricted by band pass filter.
  So that bandwidth is,

  $\boxed{B = (1 + r) R}$

  ∵ r is factor related to the filter characteristics and its value between 0 and 1.
- **Advantages of ASK :**
  1. Simplicity.
  2. Easy generation and detection.
- **Disadvantages of ASK :**
  1. It is very sensitive to noise so that it has limited applications in data communication.
  2. It is used at very low bit rates.
- **Application :**
  For data transmission at low bit rate.

1.6.3 Comparison between AM and ASK

Sr. No.	Parameter	AM	ASK
1.	Nature of input (modulating) signal	Analog	Digital
2.	Modulated signal shape	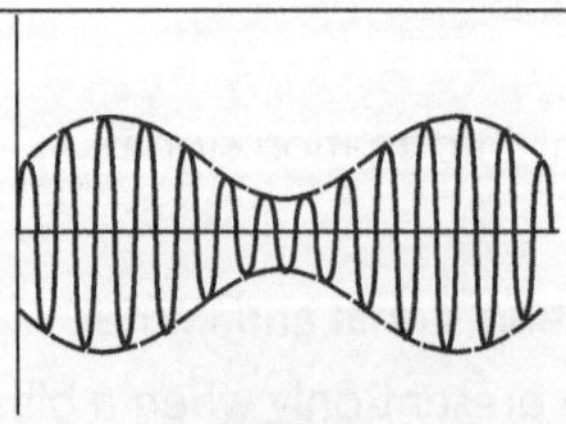 Fig. 1.19	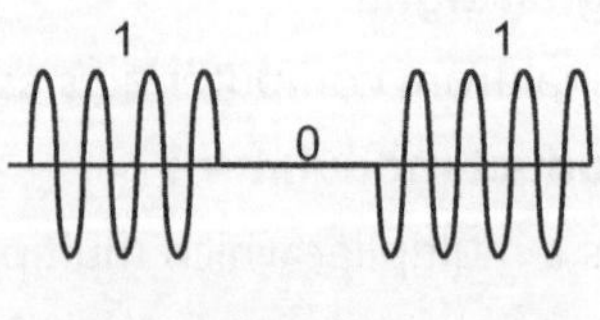 Fig. 1.20

Sr. No.	Parameter	AM	ASK
3.	Variation in carrier amplitude	Continuous variation in accordance with amplitude of modulating signal.	Carrier ON or OFF in accordance with 1 or 0 to be transmitted.
4.	Bandwidth	$2f_m$	(1 + r) R
5.	Application	Radio broadcasting	Data transmission at low bit rate.
6.	Noise immunity	Poor	Poor
7.	Number of sidebands	Two	Two
8.	Detection	Envelope	Envelope

1.7 FREQUENCY SHIFT KEYING (FSK)

- In **Frequency Shifting Keying (FSK)**, the frequency of a **sinusoidal carrier is shifted between two discrete values** where amplitude and phase remains constant.

1.7.1 FSK Transmitter

- Fig. 1.21 shows block diagram of generation of FSK.

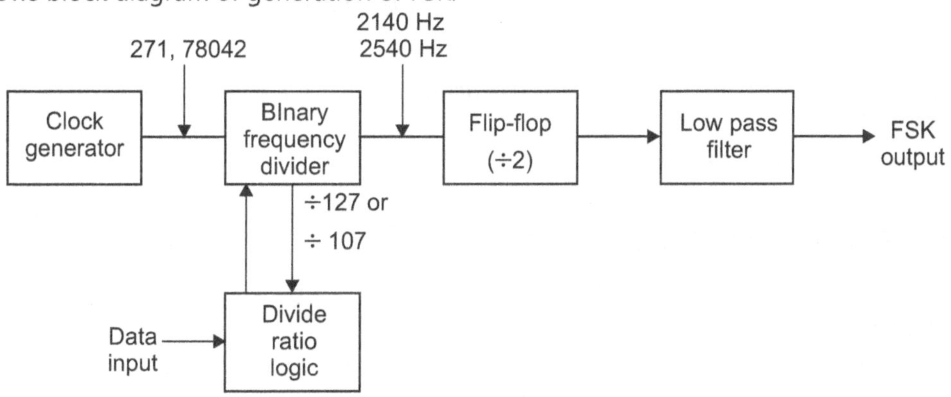

Fig. 1.21 (a) : FSK Transmitter

Note that binary 1 corresponds to frequency 1270 Hz and binary 0 to frequency 1070 Hz.

As shown in block diagram,

Clock generator :

Generates frequency of 27178042 Hz.

- **Divide ratio logic :**

  Produces frequency division by 127.

- **Frequency divider :**

  When data input is zero, the frequency divider output will be $\frac{1}{127}$ of its input. Then output frequency will be 2140 Hz.

- **Flip-flop :**

  This divides the 2140 Hz frequency by 2, producing the desired 1070 Hz output corresponding to binary '0'. Similarly, we get 1270 Hz frequency at binary '1' in which frequency divider will divide 107.

- **Low pass filter :**

  Removes higher frequency harmonics producing sinewave output.

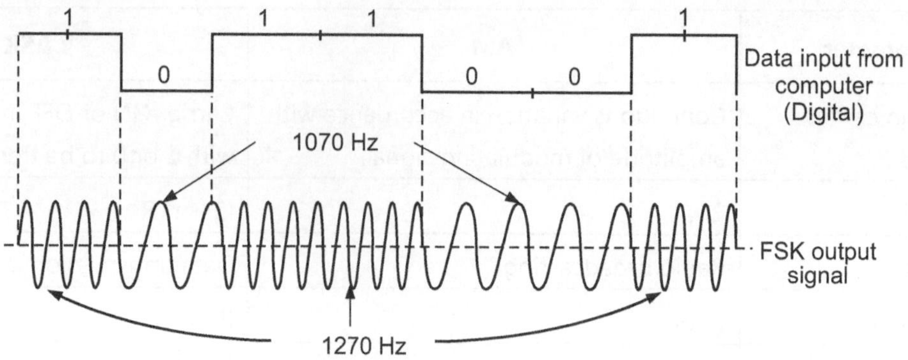

Fig. 1.21 (b) : FSK Input-Output Waveforms

- As shown in Fig. 1.21 (b), FSK signal shifts between two frequencies 1270 Hz and 1070 Hz corresponding to binary 1 and 0.

Bandwidth of FSK Signal :

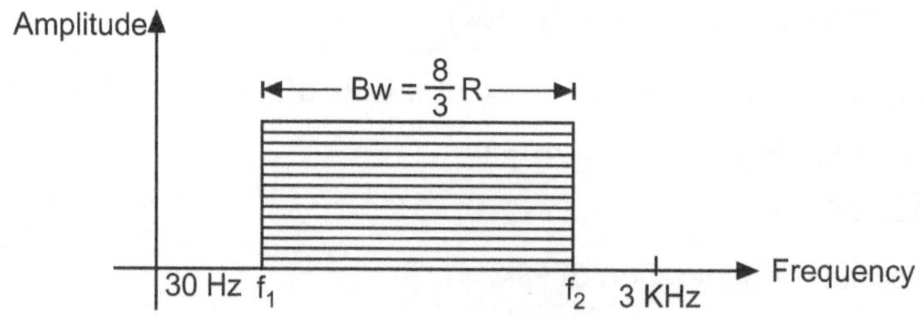

Fig. 1.21 (c)

- Both frequencies are within the 300 Hz to 3 kHz bandwidth normally associated with telephone system.

$BW = \frac{8}{3} R$ where, R is bit rate.

1.7.2 Advantages and Disadvantages of FSK

- **Advantages :**
  1. Easy to implement.
  2. Better noise immunity than ASK.
- **Disadvantages :**
  1. Not useful for high speed modems.
  2. Preferred only for low speed modems.
- **Application :**

  FSK is used in low-speed modems capable of transmitting data upto a speed 300 bit/sec.
  - This is relatively slow data rate and rarely used today.
  - Most early telephone-line modems use audio FSK to send analog receiver data upto about 300 bits/second.

1.8 BINARY FREQUENCY SHIFT KEYING (BFSK)

- In binary frequency shift keying, the frequency of the carrier is shifted according to the binary symbol.
- Whereas phase remains constant.
- i.e. there are two different frequency signals according to binary symbols.
- Let there be a frequency shift by Ω.

  Then,

$$\text{If } b(t) = 1, \; S_H(t) = \sqrt{2P_s} \cos(2\pi f_o + \Omega)\, t \qquad \ldots (1)$$

$$\text{If } b(t) = 0, \; S_L(t) = \sqrt{2P_s} \cos(2\pi f_o - \Omega)\, t \qquad \ldots (2)$$

So that, there is increase or decrease in frequency by Ω.

Table 3.1 : Conversion Table for BFSK Representation

b(t) Input	d(t)	$P_H(t)$	$P_L(t)$
1	+1	+1	0
0	−1	0	+1

∵ b(t) → Data bit to be transmitted.

P_H (t) and P_L(t) → Inputs to multiplier.

d(t) → Data bit.

We can write equation (1) and equation (2) combinely,

$$S(t) = \sqrt{2P_s} \cos [2\pi f_o + d(t)\, \Omega\, t] \quad \ldots (3)$$

∴ When symbol '1' is transmitted,

$P_H = 1$ and $P_L = 0$

The carrier frequency is $f_o + \left(\frac{\Omega}{2\pi}\right)$.

If symbol '0' is transmitted,

$P_H = 0$ and $P_L = 1$.

The carrier frequency is $-\left(\frac{\Omega}{2\pi}\right)$.

i.e.

$$f_H = f_o + \frac{\Omega}{\pi} \quad \text{For symbol '1'} \ldots (4)$$

$$f_L = f_o - \frac{\Omega}{\pi} \quad \text{For symbol '0'} \ldots (5)$$

1.8.1 BFSK Transmitter

- Fig. 1.22 shows the block diagram of BFSK transmitter.

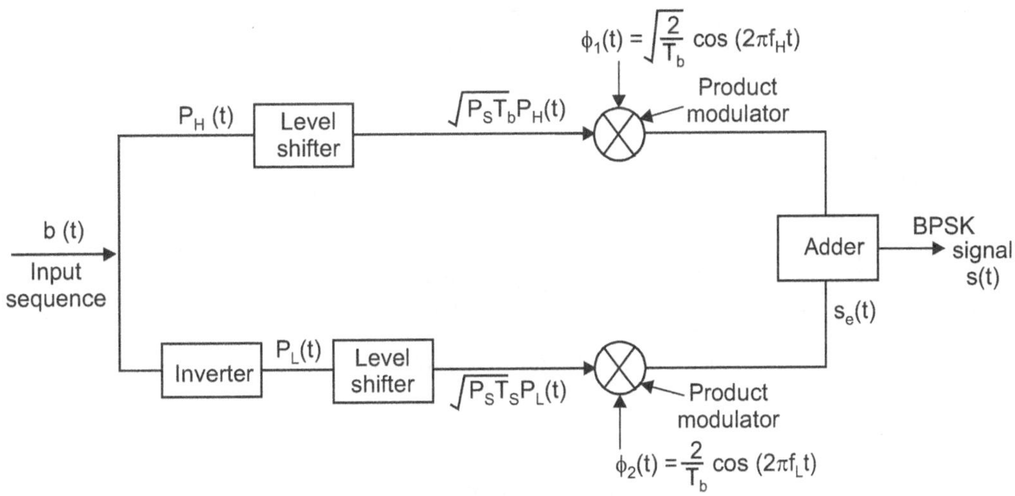

Fig. 1.22 : BFSK Transmitter Block Diagram

- We know that the input sequence b(t) is same as $P_H(t)$.

- **Inverter :**

  Inverter is added after b(t) to get $P_L(t)$.

- **Level Shifter :**
  - The level shifter input $P_H(t)$ and $P_L(t)$ are unipolar signals.
  - The level shifter converts the '+1' level to $\sqrt{P_s T_b}$, zero level is not affected.
  - Thus, the output of the level shifter will be either $\sqrt{P_s T_b}$ (if '+1') or zero (if input is zero)'.

- **Product Modulator :**
  - The two carrier signals are $\phi_1(t)$ and $\phi_2(t)$, are orthogonal to each other.
  - In one bit period of input signal (i.e. T_b) $\phi_1(t)$ or $\phi_2(t)$ have integral number of cycles.
  - Therefore, the modulated signal has continuous phase.
- **Adder :**

  The adder adds the two signals, input output signals.
  - BFSK output is shown in Fig. 1.23.

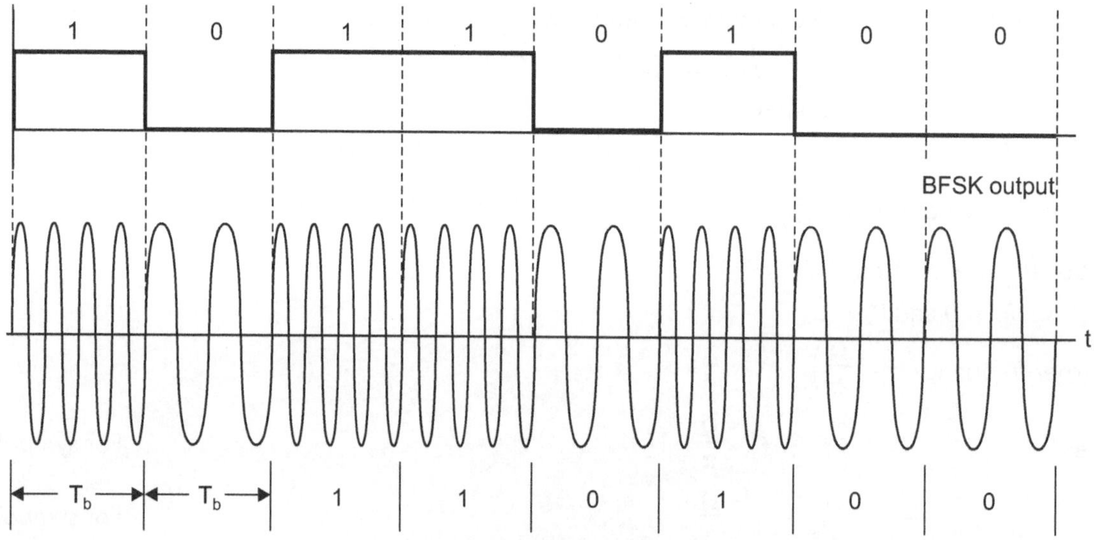

Fig. 1.23 : BFSK Signal

1.8.2 BFSK Receiver

- Fig. 1.24 shows the block diagram of BFSK receiver.

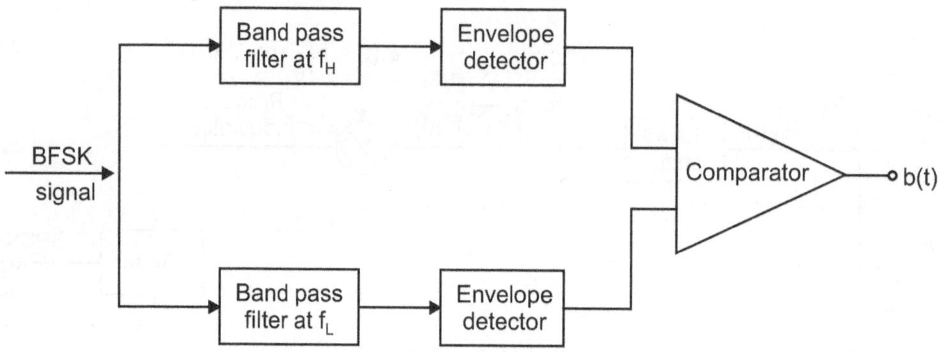

Fig. 1.24 : BFSK Receiver Block Diagram

- **Band Pass Filter :**
  - Two band pass filters, one with centre frequency f_H and other with centre frequency f_L.
  - Since $f_H - f_L = 2f_b$, the outputs of filters does not overlap.
  - The bandpass filters pass corresponding main lobes without distortion.
- **Envelope Detectors :**

  Demodulate or detects the BFSK signal into original signal.
- **Comparator :**

  The outputs of detectors are compared by comparator. If the unipolar comparator is used, then the output of comparator is bit sequence b(t).

1.8.3 Bandwidth of BFSK

Fig. 1.25 shows the power spectral density of the BFSK signal.

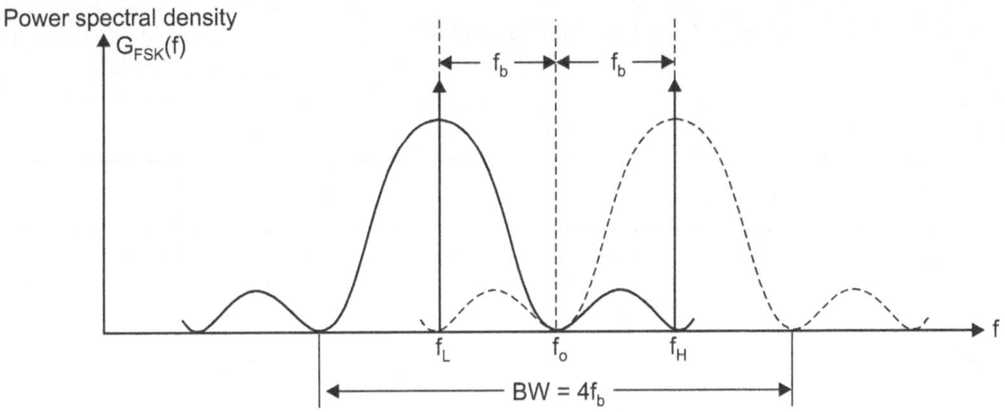

Fig. 1.25 : Power Spectral Density of BFSK Signal

- From Fig. 1.25, it is clear that the width of one lobe is $2f_b$.
- The two main lobes due to f_H and f_L are placed such that the total width due to both main lobes is $4f_b$.

$$\therefore \quad \text{BW of BFSK} = 2f_b + 2f_b$$

$$\boxed{\text{BW} = 4f_b} \quad \ldots (6)$$

Compare this bandwidth with BPSK, then,

$$\textbf{BW (BFSK)} = \mathbf{2 \times BW (BPSK)}$$

Advantages of BFSK :

Generation is easy.

Disadvantages of BFSK :

1. Greater bandwidth, which is almost double the bandwidth of BPSK.
2. Error rate of BFSK is more compare to BPSK.

1.9 PHASE SHIFT KEYING (PSK)

- In **Phase Shift Keying, the phase of carrier is varied (or shifted) to represent binary 1 and 0.**
  Both amplitude and frequency remains constant as the phase changes.
- Simplest form of PSK is called binary PSK (BPSK).
- **Binary Phase Shift Keying (BPSK)**
  - In BPSK, binary symbols '1' and '0' modulate the phase of the carrier.
    Let carrier is,

$$S(t) = A \cos (2\pi f_o t) \quad \ldots (7)$$

$$\because \quad A \rightarrow \text{Peak value of sinusoidal carrier}$$

Then, power dissipated will be,

$$P = \frac{1}{2} A^2 \quad \ldots (8)$$

$$A = \sqrt{2P}$$

When the symbol is changed, the phase of the carrier is changed by 180° or π radians.

BPSK signal is represented as,

$$S(t) = b(t) \sqrt{2P} \cos (2\pi f_o t) \quad \ldots (9)$$

Here, $b(t) = +1$ when binary '1' is transmitted

$= -1$ when binary '0' is transmitted.

- Fig. 1.26 shows binary signal and its equivalent signal b(t).

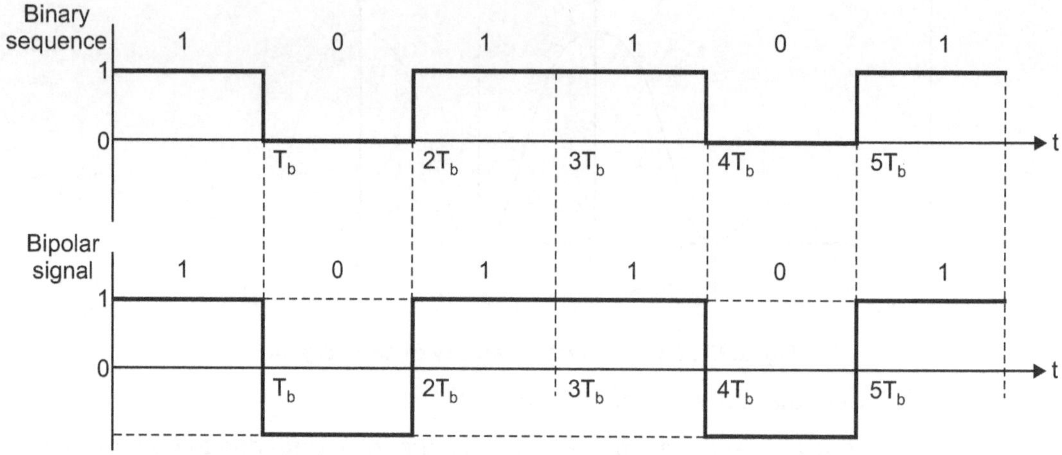

Fig. 1.26 : (a) Binary Sequence, (b) Its Equivalent Bipolar Signal b(t)

- From Fig. 3.10 (b), the signal b(t) is NRZ bipolar signal.

1.9.1 BPSK Transmitter

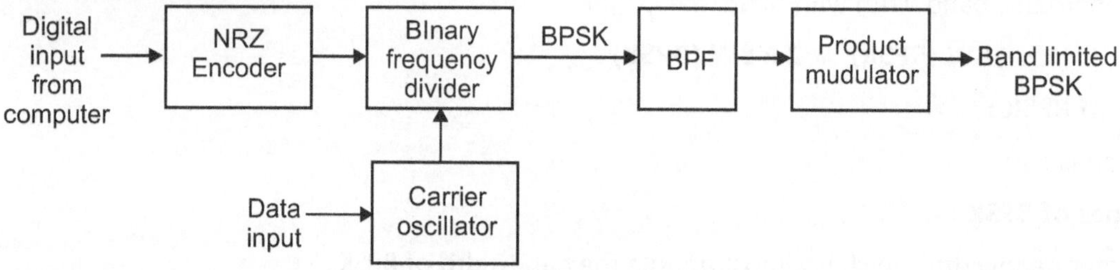

Fig. 1.27 : BPSK Generation

- **NRZ encoder :**

  This converts binary data signal (0's and 1's) into NRZ bipolar signal.

- **Carrier oscillator :**

  Generates sinewave carrier signal.

- **Product modulator :**

  Multiplies input data and carrier which results BPSK signal.

BPF :

It is bandpass filter which limits the frequency band of BPSK.

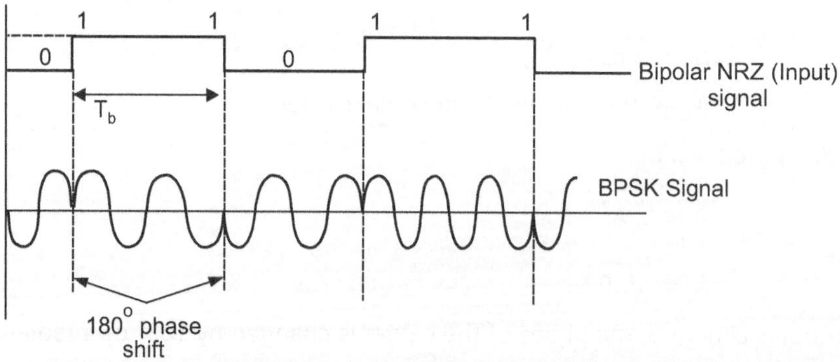

Fig. 1.28 : BPSK Input-Output Waveforms

As shown in waveform, the carrier phase is changed between 0° and 180° by the bipolar digital signal.

1.9.2 BPSK Receiver

Fig. 1.29 shows the block diagram of BPSK receiver.

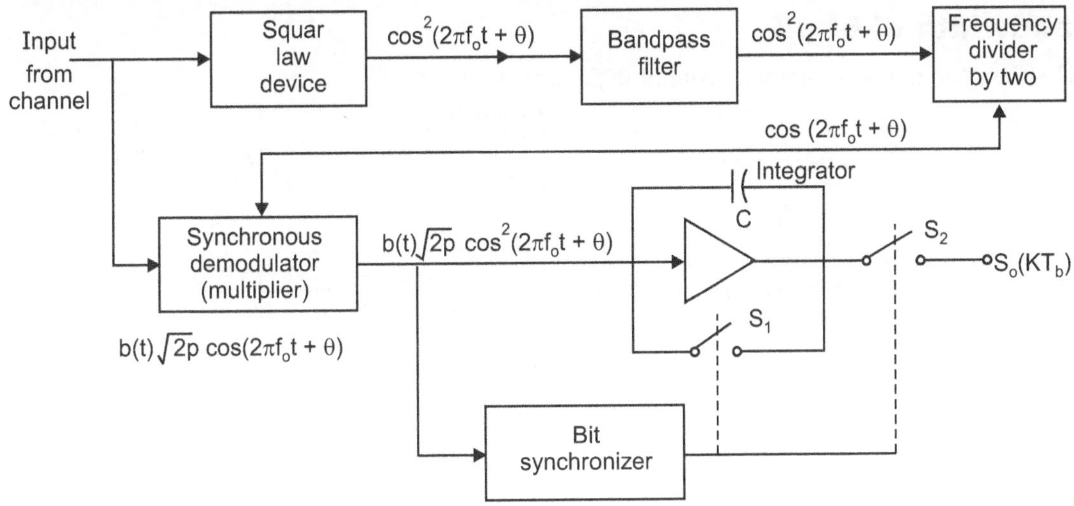

Fig. 1.29 : BPSK Receiver Block Diagram

- The transmitted BPSK signal is,

$$S(t) = b(t)\sqrt{2P}\cos(2\pi f_o t)$$

- This phase of this signal changes depending on time delay from transmitter to receiver.
- This phase change is generally fixed in the transmitted signal.
- Let phase shift is θ.

  Therefore, the signal at the input of receiver is,

$$S(t) = b(t)\sqrt{2p}\cos(2\pi f_o t + \theta) \quad \dots (10)$$

- **Square Law Device :**

  A carrier is separated from the received signal since it is coherent detection.

  Output of square law device is,

  $\cos^2(2\pi f_o t + \theta)$

- **Synchronous Demodulator :**

  This multiplies the input signal and the recovered carrier.

- **Bit Synchronizer :**

  It takes care of starting and ending times of a bit.

  - At the end of bit duration T_b, the bit synchronizer closes switch S_2 temporarily, this connects the output of an integrator to the decision device.
  - The synchronizer then opens switch S_2 and S_1 is closed temporarily, this results the integrator voltage to zero.

Integrator :

The integrator then integrates the next bit.

The output of integrator is original signal.

- One bit period T_b contains integral number of cycles of carrier.
- This means the phase change occurs in the carrier only at zero crossing. This is shown in BPSK waveforms. Fig. 1.30.

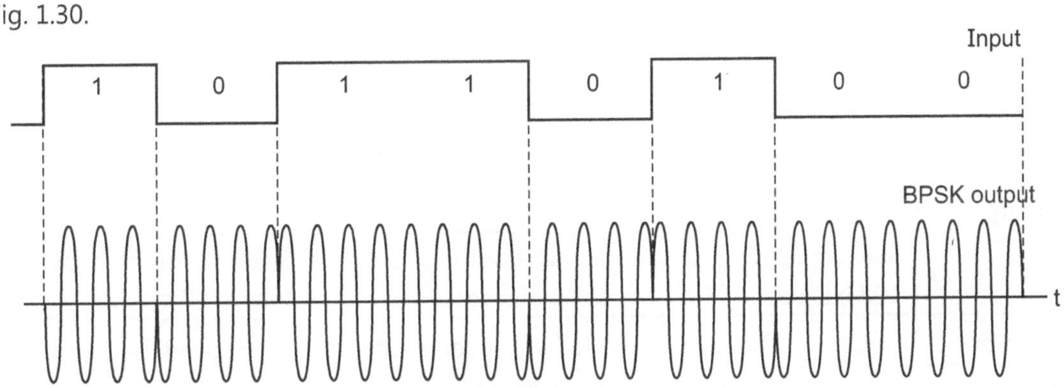

Fig. 1.30 : BPSK Waveform

1.9.3 Bandwidth of BPSK

Fig. 1.31 shows the power spectral density of BPSK signal.

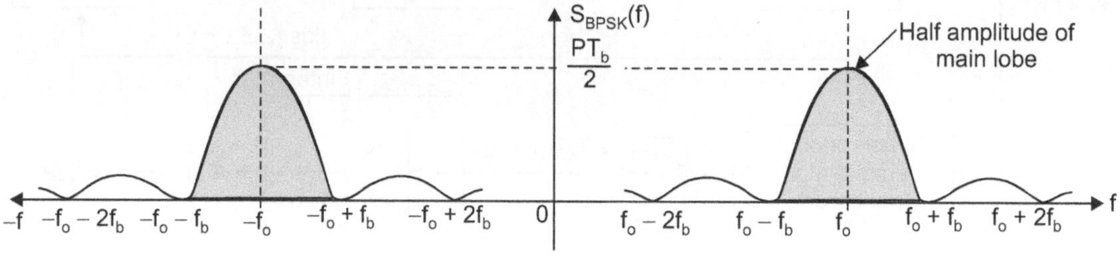

Fig. 1.31 : Plot of Power Spectral Density of BPSK Signal

- The spectrum of BPSK signal is centered around the carrier frequency f_o.

  If $f_b = \frac{1}{T_b}$

- Then, for BPSK the max-frequency in the baseband signal will be f_b shown in Fig. 1.31.

  Here lobe is centered around carrier frequency f_o and extends from $f_o - f_b$ to $f_o + f_b$.

∴ Bandwidth of BPSK signal is,

$$BW = \text{Highest frequency} - \text{Lowest frequency in the main lobe}$$

$$= f_o + f_b - (f_o - f_b)$$

$$\boxed{BW = 2f_b}$$

Hence, minimum bandwidth of BPSK signal is equal to twice the highest frequency of contained in the baseband signal.

- **Drawback :**
  - In BPSK, it is not possible to determine whether the received signal is equal to b(t) or – b(t). This results error in the output signal.
  - This problem is removed in DPSK.

1.9.4 Advantages of BPSK

1. Bandwidth less than FSK signal.
2. Good noise immunity.
3. Used for high bit rate than 1800 bits/sec.

- **Disadvantage of BPSK :**

  Generation and detection of BPSK is not easy.

- **Application of BPSK :**

  Due to low bandwidth requirement used for high speed modems.

1.10 COMPARISON BETWEEN ASK, FSK AND PSK

Sr. No.	Parameter	ASK	FSK	BPSK
1.	Variable parameter	Amplitude	Frequency	Phase
2.	Noise immunity	Low	High	High
3.	Bit rate	Suitable upto 100 bits/sec.	Suitable upto 1200 bits/sec.	Suitable for high bit rate greater than 1800 bits/sec.
4.	Application	Low speed modem	Medium speed modem	High speed modem
5.	Waveform	Refer Fig. 1.18 (b).	Fig. 1.21 (b).	Fig. 1.28.

1.11 ADVANTAGES/DISADVANTAGES OF FIBER OPTIC COMMUNICATION

The advantages of optical fiber cable over electrical cable are as follows –

(1) Greater bandwidth:

As carrier frequency is higher, the bandwidth is 106 times greater.

(2) Low loss:

Loss in co-axial cable and twisted pair wire increases with frequency whereas loss in the optical cable remains flat over a very wide range of frequencies. i.e. loss in optical fiber is slow.

(3) Electromagnetic immunity:

Glass fiber neither pick-up nor generate Electromagnetic Interference (EMI) like co-axial cables, because glass is an insulator and optical fiber has cladding. (outer coverings).

(4) Light weight:

A glass fiber weights less than a copper conductor because the copper requires more lines than fiber.

A coaxial cable weighs nine times greater than fiber. (In aircraft and automobiles).

(5) Small size:

A single fiber is capable of replacing a very large bundle of individual copper wires. For example, in telephone cable, 1000 pairs of copper wire equals the single fiber optical cable capable of handling same amount of signal.

(6) Safety:

A fiber is a dielectric. It does not carry electricity. It presents no spark or fire hazards, it does not attract lightening. It is possible to run fiber directly through a fuel tank.

(7) Security:

Fiber optic is a highly secure transmission medium. It does not radiate energy that can be received by a nearby antenna, and it is difficult to tap a fiber.

- **Disadvantages:**

There are some disadvantages of fiber optic cable.

1. Its small size and brittleness make its difficult to work with.
2. Special expensive tools and techniques required.
3. High cost.

1.12 BLOCK DIAGRAM OF FIBER OPTIC COMMUNICATION

- A block schematic of general fiber optic communication system is shown in Fig. 1.32.

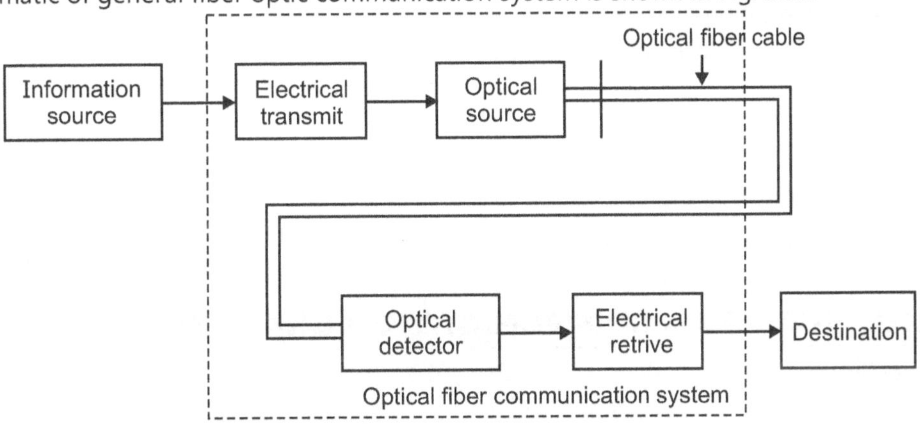

Fig. 1.32: Optical Fiber Communication System

Functions of Each Block:

Information Source:

The original information such as sound, data, picture is converted into electrical signal.

Electrical Transmit:

It drives the optical source.

Optical Source:

Converts electrical signal into light using laser or LED.

Optical Fiber Cable:

It is transmission media to travel light beam.

Optical Detector:

At receiver side optical detector converts light beam into an electrical signal.

Electrical Retrive:

Provides demodulation of optical carrier.

Destination:

Again it is original information.

1.13 MATERIALS USED FOR OPTICAL FIBRES AND TYPES

Two types of optical fiber available as regards to material used.

(1) Glass Fibers, (2) Plastic Fibers.

(1) Glass Fiber:

Glass is not a particular material, but it is a super cooled liquid because glass is a solid state of matter in which atoms are not in regular array.

- When a liquid is cooled down very rapidly, then at temperature below the freezing temperature, the liquid gets solidify in glassy or non-crystalline state as shown in Fig. 1.33.
- The material used for making fibers is silicon dioxide (silica).

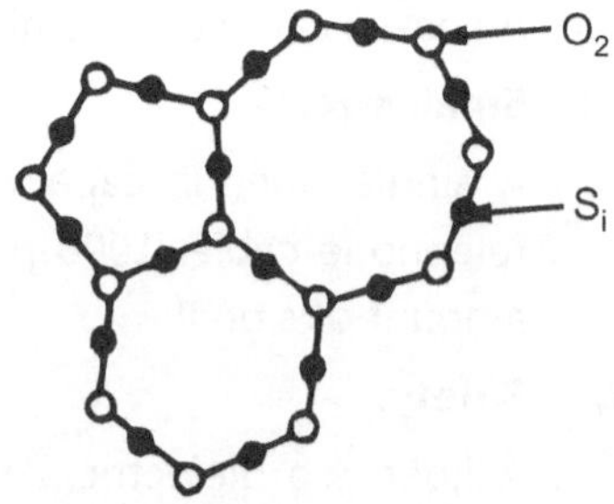

Fig. 1.33

- The silicon dioxide is mixed with oxides of genermanium boron, phosphorous to adjust its optical density slightly from that of silica. Because there are different layers of glass having different refractive indices.

(2) Plastic Fiber:

- The plastic material used for making fibers is polystyrene or poly-vinyl chloride.
- These fibers has higher losses and low withstanding capacity for temperatures.

1.14 APPLICATIONS OF FOC

1. Local and long distance telephone systems.
2. TV studio-to-transmitter interconnection, eliminating microwave radio link.
3. Closed-circuit TV systems used in buildings for security.
4. Secure communications systems at military bases.
5. Shipboard communications.
6. Computer networks, wide area and local area.
7. Aircraft controls.
8. Aircraft communications.
9. Data acquisition and control signal communications in industrial process control system.
10. Nuclear plant instrumentation.
11. Utilities (electric, gas etc.) station communications.
12. College campus communications.

1.15 OPTICAL FIBER AS A COMMUNICATION CHANNEL

Why to use Optical Fibers ?

Reasons are:

1. The light weight and non-corrosiveness of fiber make it very practical for aircraft and automotive applications.
2. A single fiber can handle as many voice channels as a 1500 pair cable can.
3. The spacing of repeaters from 35 to 80 km for fibers as opposed to from 1 to 1 $\frac{1}{2}$ km for wire is great advantage.

4. Fiber is immune to interference from lightning, cross-talk and electromagnetic radiation.
   - In order to plan the use of optical fibers in a variety of line communication applications, it is necessary to consider the various optical fibers currently available.
   - The performance characteristics of the various fiber types vary considerably depending upon the materials used in the fabrication process and the preparation technique involved.
   - In particular, high performance silica-based fibers for operation in three major wavelength regions 0.8 to 0.9, 1.3 and 1.55 μm are now widely commercially available.
   - Complex refractive index profile single-mode fibers including dispersion modified fibers and polarization maintaining fibers are also commercially available and in the former case are starting to find system application within communications.
   - Another relatively new area of commercial fiber development is concerned with mid-infrared wavelength range 2 to 5 μm, often employed heavy metal fluoride glass technology.
   - However, the fiber products that exists for this wavelength region tend to multimode with relatively high losses and hence at present are only appropriate for special applications.

1.16 CONSTRUCTION OF FIBER OPTIC CABLE

- The fiber, which is called the core, is usually surrounded by a protective cladding shown in Fig. 1.34.
- The cladding is also made of glass or plastic but has lower index of refraction.
- This ensures that proper interface is achieved so that light waves remain within the core.
- In addition to protecting the fiber core from nicks and scratches, the cladding adds strength.

  [Some fiber-optic cables have a glass core with a plastic cladding].

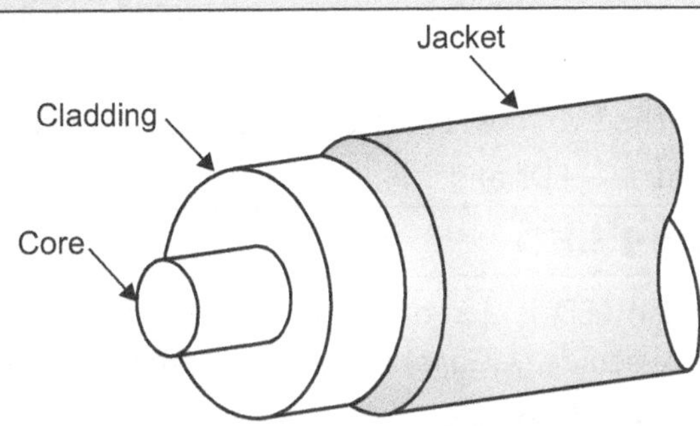

Fig. 1.34: Basic Construction of Fiber Optic Cable

- Others have a plastic core with a plastic cladding.
- Another common arrangement is a glass core with a plastic cladding. It is called plastic clad silica (PCS) cable.
- A plastic jacket as a insulation is put over the clad.

1.17 OPTICAL FIBER TYPES AND CHARACTERISTICS

The characteristics of light transmission through a glass fiber depends on many factors such as:

1. The composition of the fiber.
2. The amount and type of light introduced into the fiber.
3. The diameter and length of the fiber.

- The composition of the fiber determines the refractive index.
- By a process called doping, in which other materials are introduced into the material that change its index number. This process produces a single fiber with a core index n_1 and a surface index (cladding) n_2 typically $n_1 = 1.48$ and $n_2 = 1.46$.
- Another characteristics of the fiber, which depends on its size, is its **mode of operation.**
- The term **mode** is used refers to mathematical and physical descriptions of the propagation of energy through a medium.

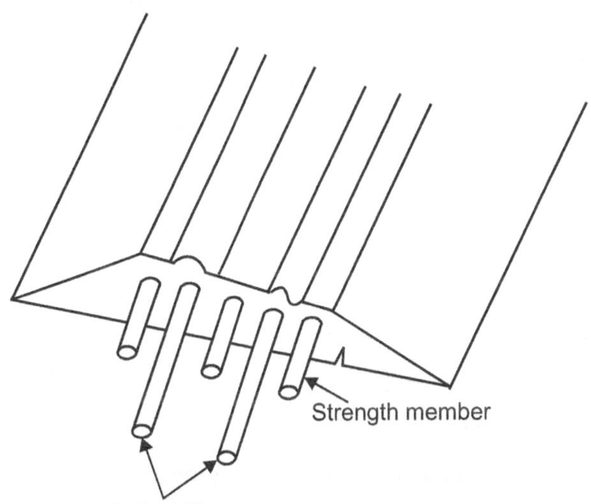

Fig. 1.35: Fiber Cable Assembly

- The number of modes supported by a single fiber can be as low as 1 or as high as 1,00,000 so that fiber can provide a path for one light ray or for hundreds of thousands of light rays.
- From this characteristics introduce the terms **single mode** and **multimode.**
- Another term which is the refractive index profile.
- It describes the relationship between the multiple indices which exist in the core and cladding of the particular fiber i.e. **"light changes speed when it passes from one medium to another".**

1.18 OPTICAL SOURCES/TRANSMITTERS

- In an optical communication systems, transmission begins with the transmitter, which consists of a modulator and the circuitary that generates the carrier.
- In optical communication, carrier is a light beam that is modulated by digital pulses which turn in ON and OFF.
- **The basic transmitter is known as source.**
- **Conventional light sources such as incandescent lamps cannot be used in fiber optic systems** because they are simply too slow.
- An incandescent light source consists of a filament that heat up and emits light.
- Such a light source cannot be turned OFF and ON fast enough because of the thermal delay in the filament.
- So that, in order to transmit high-speed digital pulses, a very fast light source must be used.
- The two most commonly used light (optical) sources are:

  1. LEDs and 2. LASERs.

1.19 LED

- An LED is a p-n junction diode usually made from a semiconductor material such as aluminium-gallium-arsenide (AlGaAs) or gallium-arsenide-phosphide (GaAsP).

Working Principle:

- LEDs emit light by spontaneous emission, light emitted as a result of the recombination of electrons and holes.
- When current is passes through the PN structure of LED, electrons are injected into an active layer. Light is generated when these injected electrons return from their excited (higher) energy levels to their normal (lower) energy levels. The energy difference is releases as light.

Construction:

- The construction of the LED is shown in Fig. 1.36.
- In the Fig. 1.36, the surface layer is p–type.
- A shallow p-n junction is formed and electrical contacts made to both regions.
- The upper part of the p-material is left uncovered so that radiations from the device are not much affected.
- The inner quantum efficiency of the LED material, sometimes is nearly 100%.
- The external efficiency is much lower.

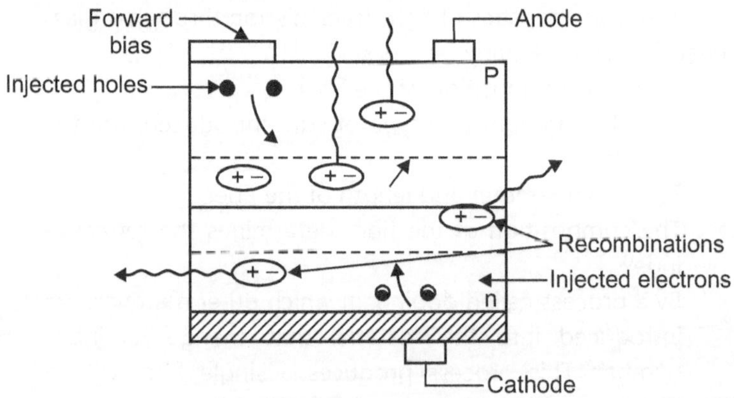

Fig. 1.36: LED Construction

- This is so since most of the emitted radiation strikes the material interface at an angle which is greater than the critical angle and hence, gets trapped.
- The semiconductor/air interface in the LED has the shape of a hemisphere.
- This is made so such that most of the rays strike the surface at less than the critical angle.

Working:

- When forward biased, minority carriers are injected across the P-N junction.
- Once across the junction, these minority carriers recombine with majority carriers and give up energy in the form of light.

Types of LED:

1. Surface emitter LED.
2. Edge emitter LED.
3. Super luminescent LED.
4. Planar LED.
5. Dome LED.

Advantages:

As already stated, LED has a number of advantages. They have –

(a) very high speed (in n sec)
(b) different colours
(c) long life
(d) low voltage operation
(e) low cost
(f) linearity of radiant power output with forward current over a wide range.
(g) low voltage operation to make it compatible with ICs.

Disadvantages:

However, it has some outstanding disadvantages.

(a) It is sensitive to over voltage and over current due to which it is easily damaged.
(b) The radiant output power and wavelength are temperature dependent.
(c) It is difficult to get the theoretical overall efficiency except by using special cooled or pulse conditions.
(d) Large power dissipation.

Applications:

1. They are used in 7-segment, 16-segment and dot-matrix displays, which are used to indicate alphanumeric characters and symbols.
2. They are used for indicating power ON/OFF conditions.
3. They are used in burglar alarm systems.
4. They are used to indicate digital logic state.
5. They can be used as solid state video displays in place of CRTs.
6. They are used in image sensing circuits in picture phones.

Colours:

- By using different LED materials and different dopants, it is possible to get different colours.
- LEDs may be made from gallium phosphide (GaP), gallium arsenide (GaAs) and silicon carbide (SiC). SiC gives yellow light, gallium arsenide phosphide gives red light while gallium arsenide gives infrared radiation. GaP with nitrogen gives green light.

1.19.1 Edge Emitter LED

- The edge emitting LED which was developed by RCA, shown in Fig. 1.37.
- These LEDs emit a more directional light pattern than surface emitting LED.

Construction:

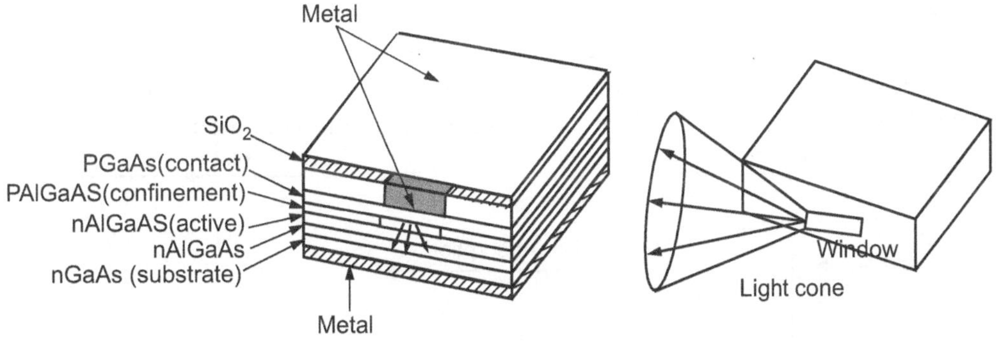

(a) (b)

Fig. 1.37: Edge Emitting LED

- The construction is similar to the planar and surface emitting diodes except that the emitting surface is stripe rather than a confined circular area.
- The light is emitted from an active stripe and forms an elliptical beam.
- The radiant light power emitted from an LED is a linear function of the forward current passing through the device.
- It can also be seen that the optical output power of an LED is, in part, a function of the operating temperature.

1.19.2 Surface Emitting LED

- For more practical applications, such as telecommunications, data rates in excess of 100 Mps are required. For these applications etched-well surface emitting LED was developed.

Construction:

- Fig. 1.38 (c) shows surface emitting LED.
- This LED emits light in many directions.
- The etched well helps to concentrate the emitted light to a very small area.
- These devices are more efficient than the standard surface emitters, and they allow more power to be coupled into the optical fiber, but they are also more difficult to be coupled into the optical fiber.

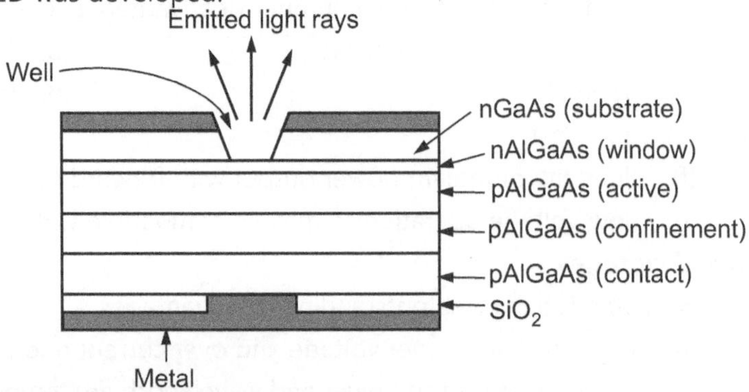

Fig. 1.38 (c): Surface Emitting LED

1.20 LASER (LIGHT AMPLIFICATION BY STIMULATED EMISSION OF RADIATION)

- The laser is a device which amplifies the light, hence the LASER is an acronym for light amplification by stimulated emission of radiation.
- The operation of the device may be described by the formation of an electromagnetic standing wave within a cavity (optical resonator) which provides an output of monochromatic highly coherent radiation.

Principle:

- **Laser action is the result of three process absorption of energy packets (photons), spontaneous emission, and stimulated emission.**
- **These processes are represented by the simple two-energy-level diagrams.**

  where, E1 is the lower state energy level.

  E2 is the higher state energy level.
- Quantum theory states that any atom exists only in certain discrete energy state, absorption or emission of light causes them to make a transition from one state to another.
- The frequency of the absorbed or emitted radiation is related to the difference in energy E between the two states.

  If E1 is lower state energy level

  and E2 is higher state energy level

  $$E = E2 - E1 = h \cdot f$$

  where, $h = 6.626 \times 10^{-34}$ Js (Planck's constant).
- An atom is initially in the lower energy state, when photon with energy E2 – E1 is incident on the atom it will be excited into the higher energy state E2 through the absorption of the photon.

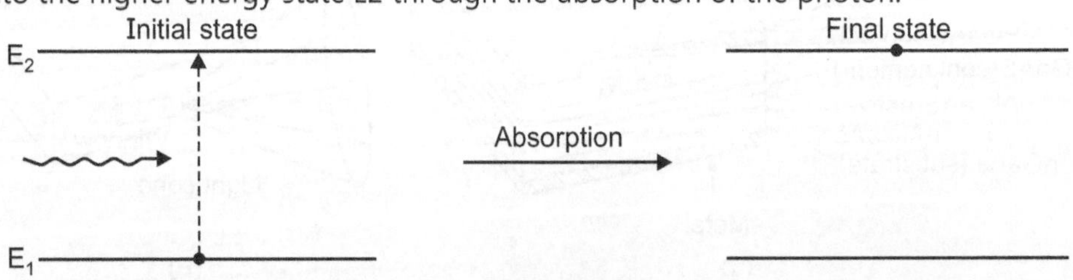

Fig. 1.39: Absorption

- When the atom is initially in the higher energy state E2, it can make a transition to the lower energy state E1 providing the emission of a photon at a frequency corresponding to E = h · f.
- The emission process can occur in two ways:
  (a) By spontaneous emission in which the atom returns to the lower energy state in random manner.
  (b) By stimulated emission when a photon having equal energy to the difference between the two states (E2 – E1) interacts with the atom causing it to the lower state the creation of the second photon.

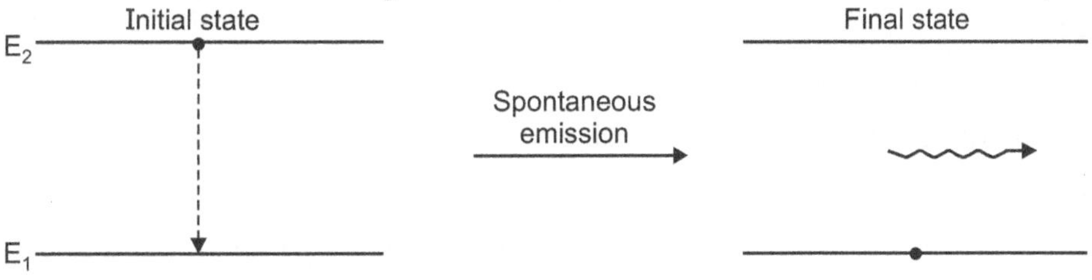

Fig. 1.40 (a): Spontaneous Emission

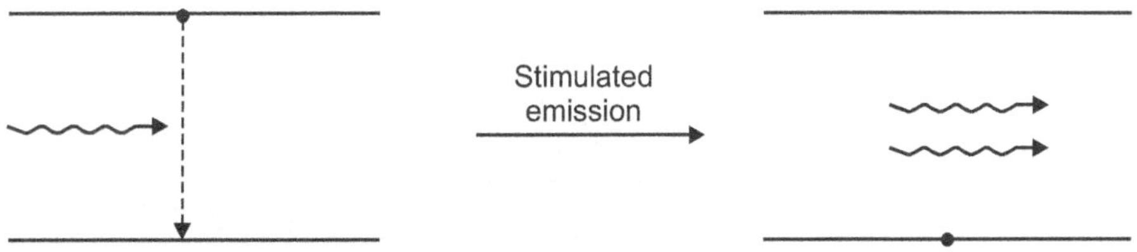

Fig. 1.40 (b): Stimulated Emission

- Spontaneous emission gives incoherent radiation.
- Stimulated emission gives coherent radiation.
- Since the light associated with emitted photon is of same frequency of incident photon, and in same phase with same polarization.
- It means that when an atom is stimulated to emit light energy by an incident wave, the liberated energy can add to the wave in constructive manner.
- The emitted light is bounced back and forth internally between two reflecting surfaces.
- The bouncing back and forth of light wave causes their intensity to reinforce and build-up.
- The result is a high brilliance, single frequency light beam providing amplification.

Construction:

- LASERs are constructed for many different materials, including gases, liquids and solids.
- The construction of LASER is similar to LED except that the ends are highly polished.
- The mirror like ends trap the photons in the active region and as they reflect back and forth, stimulate free electrons to recombine with holes at higher-than normal energy level. This is called lasing.

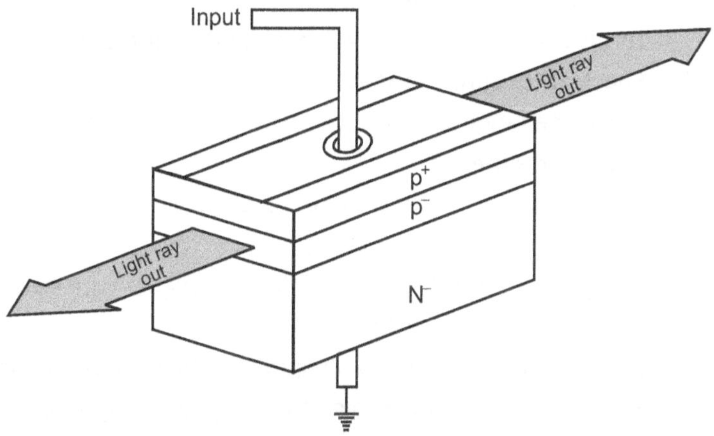

Fig. 1.41: LASER Diode Construction

Working:

- As current passes through a forward biased P-N junction diode, light is emitted by spontaneous emission at a frequency determined by the energy gap of the semiconductor material.
- When a particular current level is reached the number of minority carrier and photons produced on either side of the P-N junction reaches a level where they begin to collide with already excited minority carriers.

- This causes an increase in the ionization energy level and makes the carriers unstable.
- When this happens, a typical carrier recombine with an opposite type of carrier at an energy level that is above its normal before collision value.
- In the process, two photons are created, one is stimulated by another.
- Essentially, a gain in the number of photons is realized.
- For this to happen, a large forward current that can provide many carriers (holes and electrons) is required.

Advantages over LED:

1. Developing light power upto several watts.
2. More powerful than LEDs and therefore capable of transmitting much lower distances.
3. Ability to turn OFF and ON at a faster rate.
4. High speed laser diodes capable of gigabit per second digital data rates.

1.21 COMPARISON OF DIFFERENT SOURCES

Sr. No.	Parameter	LED	LASER
1.	Principle of operation	Spontaneous emission	Stimulated emission
2.	Spectral width	Broad (20-100 nm)	Narrow (1-5 nm)
3.	Data rate	Low	Very high
4.	Transmission distance	Smaller	Greater
5.	Temperature sensitivity	Less	More
6.	Coupling efficiency	Very low	High
7.	Compatible fibers	Multi-mode step index	Single mode step index
8.	Cost	Low	High
9.	Noise	More	Less

1.22 OPTICAL DETECTORS/RECEIVERS

Predominant types of receiver used in communication system rely on principle of ionization of semiconductor. Optical receivers are placed at the receiver side to convert light ray into electrical signal.

Four important parameters:

1. **Detector responsivity:**

   This is the ratio of output current to input optical power. Hence, this is efficiency of the device.

2. **Spectral response range:**

   This is the range of wavelength over which device will operate.

3. **Response time:**

   This is the time required for responding variation in light intensity.

4. **Noise:**

   Level of noise produced during lowest level of light detection.

- **Requirements of Detectors:**

  1. The detector should be highly sensitive at operating wavelength.
  2. Produce a maximum electrical signal strength for a given amount of optical power.
  3. Noise introduced by detector should be minimum.
  4. Stable performance characteristics.
  5. Small physical size.
  6. Less bias voltage or current requirement.
  7. Less cost.

1.23 PHOTO DIODES

- Photo diodes convert light directly to electric current.
- PIN diode can convert one photon to one electron of current. Hence, amplifier is needed with receiver.

Photo diodes

↓

PIN Photo diode | Avalanche Photo Diode

1.23.1 PIN Diodes

Name PIN because intrinsic layer introduced between P and N region.

Principle:

Principle of PIN diodes is reverse that of LED. That is, light is absorbed at p-n junction rather than emitted.

The wide intrinsic i layer has only small degree of doping and has wide depletion layer. There are number of modifications:

1. It increases the chances of entering photon being absorbed, because amount of absorbant material is more.
2. Since junctions are wider, junction capacitance is less.
3. Two ways of current flow are diffusion and drift.

Thus i layer increases responsivity and decreases response time.

Construction:

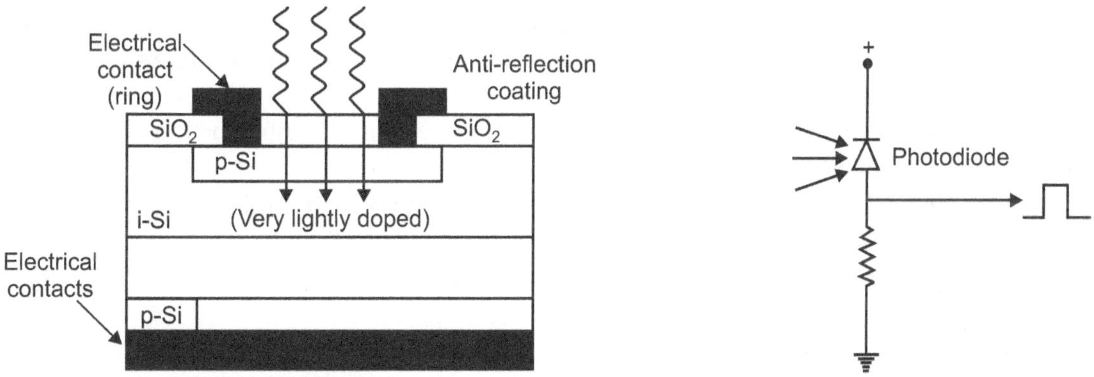

(a) Typical Silicon PIN Diode Schematic **(b) How diode converts light into voltage pulses**

Fig. 1.42

Working:

- The sensitivity and response time can be increased by adding an undoped intrinsic (I) layer between the P and N semiconductor to form a PIN diode.
- The I layer exposed to light.
- This is a silicon PN junction diode that is sensitive to light.
- This diode is normally reverse-biased.
- The only current that flows through it is an extremely small leakage current.
- Whenever light strikes the diode, this leakage current will increase significantly.
- It flows through a resistor and develops a voltage drop across it.
- The result is an output voltage pulse.
- Voltage pulse is small, so that it must be amplified.
- For operating PIN diode, energy of absorbed photon must be sufficient to promote an electron across bandgap.
- At energy higher than bandgap energy, material absorbs photons.
- Germanium is material used as it has low bandgap energy regardless of wavelength.

Characteristics of PIN diodes:

1. **Quantum efficiency:** It is defined as the ratio of number of electrons collected at the junction over the number of incident photons. PIN diodes has 1 electron release per 1 photon.
2. **Responsivity:** It is a measure that does take photon energies into account. It is output photo current of device divided by the input optical power.

Typical responsivity for λ = 900 nm is 0.44.

1.23.2 Avalanche Photo Diodes (APDs)

- APDs amplify the signal during the detection process. They use a similar principle to that of photomultiplier tubes used in nuclear detection.

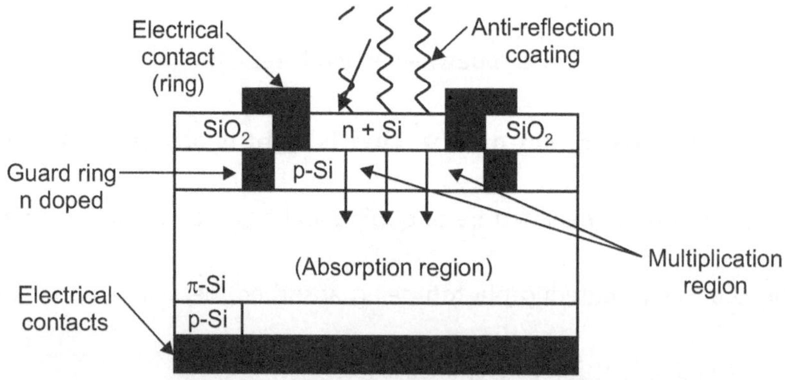

Fig. 1.43: Avalanche Photo Diode (APD)

Principle:

A single photon acting on the device releases a single electron. This electron is accelerated through an electric field until it strikes a target material.

This collision releases multiple electrons. These electrons get accelerated and strike another target.

This releases more electrons. This process is repeated. Thus, one electron has created large current.

Working:

- Structural difference between APD and p-i-n diode is that i zone (which is lightly n-doped in p-i-n structure) is lightly p-doped and renamed π layer.
- Arriving photons generally pass straight through n^+-p junction (because it is thin) and are absorbed in π layer.
- This absorption produces free electron.
- Electric potential across π layer is sufficient to attract the electron towards one contact and holes at other contact.
- Electrons are attracted towards n^+ layer at top of the device.
- Around the junction between n^+ and p-layers the electric field is so intense that carriers are strongly accelerated.
- When these electrons collide with other atom, they produce electron-hole pair.
- This process is called as impact ionization.
- When the light strikes the junction, breakdown occurs and large current flows.
- This high reverse current requires less amplification.
- Like PIN diode APD also reverse biased.
- When a sufficient amount of reverse will flow due to avalanche effect.

APD Characteristics:

1. **Sensitivity of APDs:**
   Very high sensitivity makes APD very useful.
2. **Gain bandwidth product:**
   Performance of optical detector is decided by its gain bandwidth product.
   APD has gain bandwidth product of 150 GHz.
3. **Operating speed:**
   Device capacitance limit speed of p-i-n diodes.
   APD takes too long to make impact ionization effect.
   Hence, maximum speed of APD is restricted.

Advantages:

- It is fastest and most sensitive photodiode.
- Capable to handle very high gigabit/sec. data rates possible in some systems.

Disadvantage:

- It is expensive and circuitry is complex.

Important Points

- **Multiplexing** is the set of techniques that allows the simultaneous transmission of multiple signals across a single data link.
- Multiplexer means may into one.
- Demultiplexer means one into many.

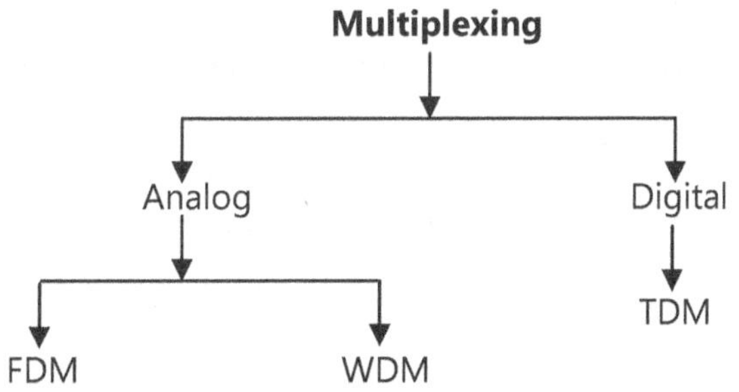

- **FDM** means total range of frequency is divided into number of frequency slots.
- **WDM** is an analog multiplexing technique to combine optical signals.
- **TDM** is a digital multiplexing technique to combine data.
- In **TDM**, the data rate of the link is n times faster and the unit duration is n times shorter.
- **Modulation** of binary data or digital to analog modulation is the process of a changing one of the character of analog signal in a digital signal.

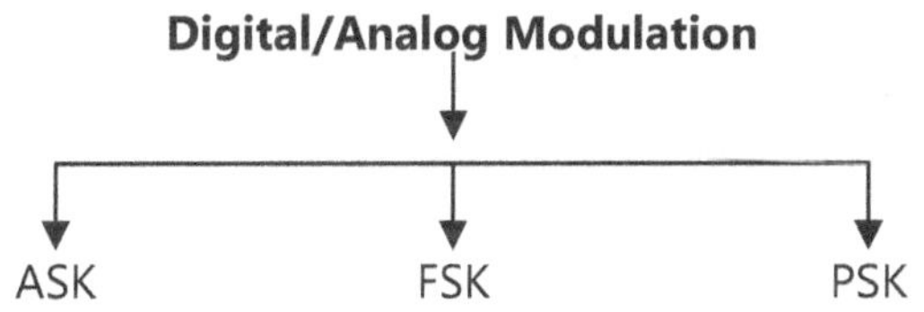

- **ASK**, the amplitude of carrier is varied to represent binary 1 and 0, both frequency and phase remains constant.
- Bandwidth of ASK $\boxed{B = (1 + r)\,R}$.
- In **FSK**, the frequency of sinusoidal carrier is shifted between two discrete values where amplitude and phase remains constant.
- In FSK, binary 1 corresponds to frequency 1270 Hz and binary 0 to frequency 1070 Hz.
- Bandwidth of FSK, $\boxed{BW = \frac{8}{3} R}$.
- Bandwidth of BFSK, $\boxed{BW = 4f_b}$.
- In **PSK**, the phase of carrier is varied or shifted to represent binary 1 and 0.
- Bandwidth of **BPSK**, is $\boxed{BW = 2f_b}$.
- A single fiber can handle as many voice channels as a 1500 pair cable can.
- Fiber is immune to interference from lightning, cross-talk and electromagnetic radiation.
- The character of light transmission through a glass fiber depends on many factors such as:
  1. The composition of fiber.
  2. The amount of type of light introduced into the fiber.
  3. The diameter and length of fiber.

- An **optical fiber** is like electrical transmission line in which light is transmitted.
- The **name optical fiber** because the signal in the form of light travels inside the glass fiber due to internal reflection.
- Because of the frequency of light is extremely high it can accommodate very wide bandwidths of information and extremely high data rates can be achieved with excellent reliability.
- Light is a kind of electromagnet radiation.
- The visible light range with infrared and ultraviolet rays is called **optical spectrum.**
- Basic elements in fiber optic system are:

  1. Information source. 2. Optical source. 3. Optical detector.
- A fiber optic cable is a glass pipe used to carry a light beam from one place to another.
- Light is a kind of electromagnetic radiation.
- This visible light range between infrared and ultraviolet rays is called optical spectrum, of the range 3 × 10" to 3 × 106 Hz.
- Thus, semiconductor optical sources (Laser and LED) and detectors (photodiodes and phototransistors) compatible in size with optical fibers were designed and fabricated to enable successful implementation of the optical fiber system.

Practice Questions

1. Define analog and digital signal.
2. Explain channel multiplexing.
3. Describe multiplexing techniques.
4. Explain FDM.
5. Explain FDM transmitter and receiver.
6. Give advantages and disadvantages of FDM.
7. Explain WDM system with advantages, disadvantages and applications.
8. Describe TDM in detail with block diagram.
9. Explain synchronous TDM.
10. Compare FDM and TDM.
11. Give modulation formats.
12. Explain ASK transmitter with bandwidth.
13. Describe FSK transmitter with waveform.
14. Give advantages and disadvantages of FSK.
15. Explain BPSK transmitter and receiver with neat block diagram.
16. Compare ASK, FSK and BPSK.
17. Give advantages and disadvantages of FOC.
18. Explain block diagram of FOC.
19. How optical fiber can work as communication channel.
20. Explain LED with construction, advantages and disadvantages.
21. Give principle of LASER.
22. Compare LED and LASER.
23. Give types of photodiodes, explain any one.

***

Chapter **2**

OPTICAL FIBRES GEOMETRICAL

2.1 INTRODUCTION

- As we consider the optical fiber in various geometrical aspects, it becomes easier to understand.
- Optical fiber is divided into two main categories: Step index and Graded index fiber.
- Wave propagation is done by considering the propagation is done by considering the propagation of light in step-index fibers using Maxwell's equation for EM waves.
- Dispersion of transmitted optical signal causes distortion for both digital and analog transmission.
- It includes group velocity dispersion, material dispersion, waveguide dispersion, polarization mode dispersion.
- The predominant losses in optical fiber are: – Absorption loss, – Scattering loss, – Dispersion loss etc.
- There are three fabrication methods that are commonly used in manufacturing process of fibers.

  1. MCVD, 2. OVD, 3. VAD.

2.2 OPTICS DESCRIPTION

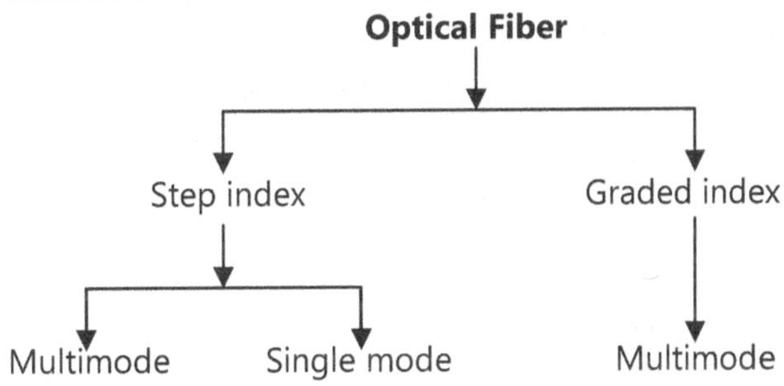

- Considering the refractive index profile and mode, optical fiber is classified as,

  1. Multimode step index fiber, 2. Multimode graded index fiber, 3. Single mode step index fiber.

2.2.1 Multi-mode Step Index Fiber

- It is also known as simply Step-Index Fiber.
- **The fiber which has uniform distribution of refractive index n_1 throughout its core is called as Step Index Fiber.**
- The cladding has uniform refractive index distribution as shown in Fig. 2.1.

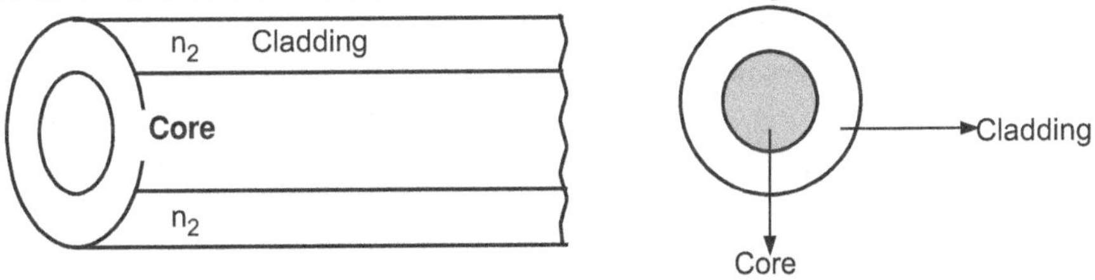

Fig. 2.1: Uniform Distribution of Refractive Index in Step Index Fiber

- Step index refers to rapid change in refractive index at the core-clad interface.
- The refractive indices of core and cladding are slightly different.
- Refractive index of core is more than that of cladding.
- From core to clad, refractive index immediately changes in step i.e. from n_1 to n_2.
- **Multimode Propagation** in Step Index consists of many light rays.

- Step Index provides reflection of light rays because of immediate change in refractive index between the core and cladding. It gives zigzag transmission of light.

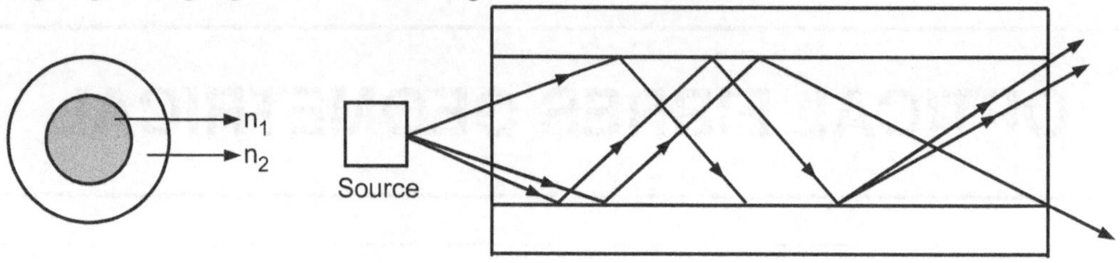

Fig. 2.2: Multimode Propagation in Step-Index Fiber

Advantages:

1. Multimode step-index fiber is of large size.
2. Typical core diameter are in 50 to 1000 μm range.
   Such large diameter cores are excellent of gathering light and transmitting it efficiently i.e. an inexpensive light source such as LED can be used to produce the light pulses.
3. It is also easiest to make.

Application:

Multimode-step-index is widely used for short to medium distances at relatively low pulse frequencies.

Limitation:

- In Fig. 2.3, a short light pulse is applied to the end of cable by the source.
- Light rays from the source will travel by multiple path.
- At the end of cable, those rays which travel the shortest distance reach the end first.
- Other rays begin to reach the end of the cable later in time until the light ray with the longest path finally reaches the end concluding the pulse.
- Ray A reaches end first, then B and then C.
- The result is a pulse at the other end of the cable that is lower in amplitude due to the attenuation of the light in the cable and increased duration due to the different arrival times of various light rays. **This stretching of the pulse is known as modal dispersion.**
- But the pulse has been stretched, input pulses cannot occur at a rate faster than the output pulse duration permits.
- Otherwise the pulses will merge shown in Fig. 2.4.
- At the output, one long pulse occurs and will be indistinguishable from the three separate pulses originally transmitted.
- i.e. incorrect information will be received.

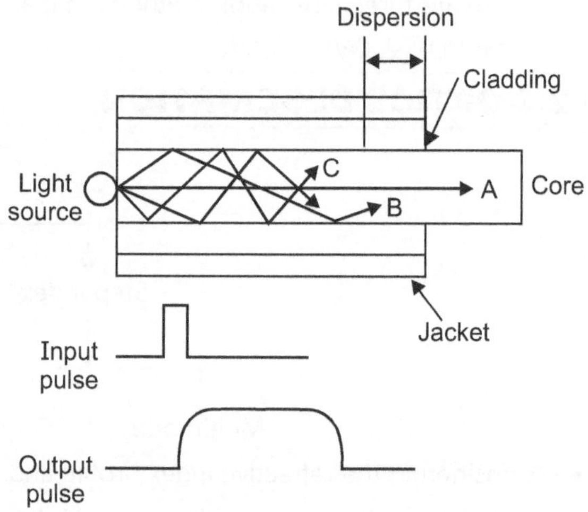

Fig. 2.3: A Multimode Step Index

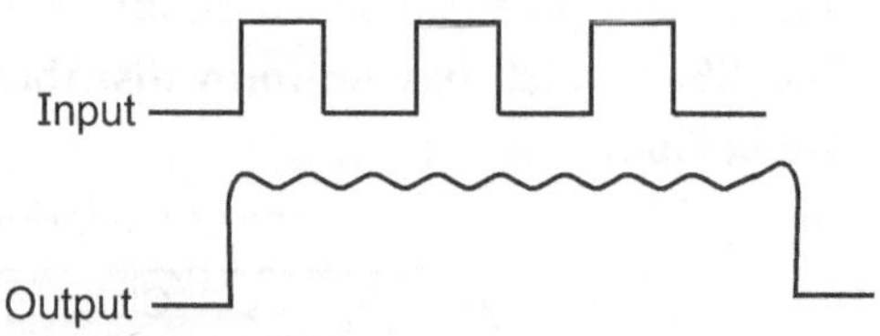

Fig. 2.4: The Effect of Modal Dispersion on Pulses occurring too rapidly in a Multimode Step Index Cable

Solution to above problem:

- The only cure for this problem is to reduce the repetition rate or the frequency of the pulses.
- When this is done, proper operation occurs.
- But with pulses at lower frequency, less information can be handled.

2.2.2 Multi-mode Graded Index Fiber

- It is also known as simply Graded Index Fiber. It has a non-uniform core.
- **The core index is highest at center and decreases until it matches that of cladding.**

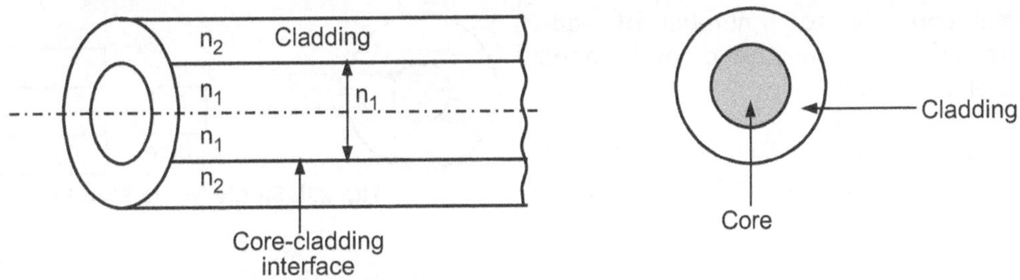

Fig. 2.5: Non-uniform core in Graded Index Fiber

- Here the refractive index of core represented as n_1.
- So that value of n_1 is maximum at center of the fiber i.e. at the axis.
- Away from center, value of refractive index decreases.
- But this minimum value of n_1, at the core-clad interface is always greater than refractive index of cladding.

Multi-mode Propagation

- In graded index, more gradual change of refractive index occurs, hence light rays are bent towards the axis of fiber in the form of refraction.
- It gives gradual bending effects of light as shown in Fig. 2.6.

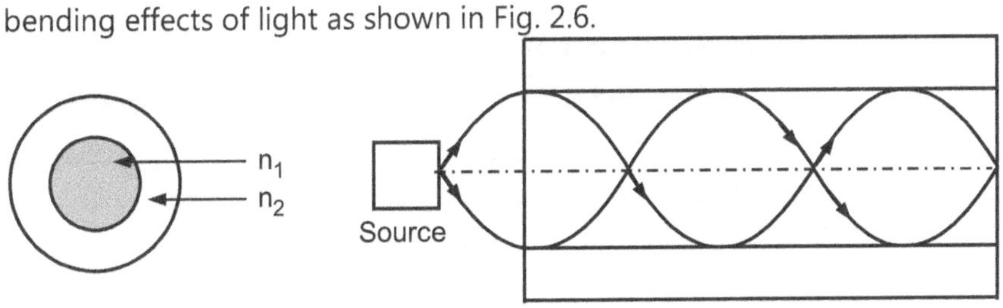

Fig. 2.6: Multi-mode propagation in Graded Index Fiber

Advantage over Multimode-Step-Index:

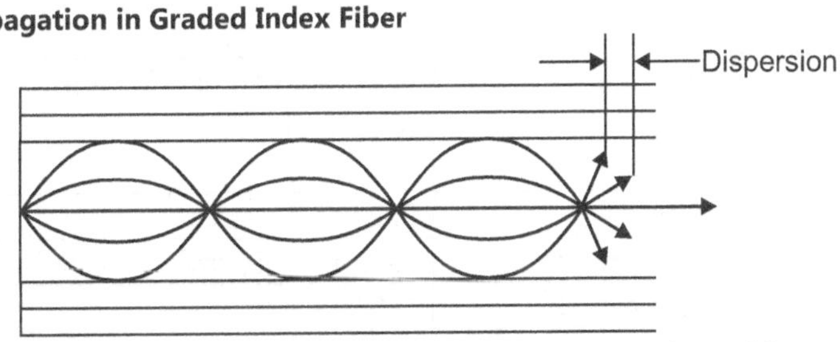

Fig. 2.7: Dispersion in Multimode Graded-Index Cable

- Fig. 2.7 shows the typical paths of the light beams.
- Because of continuously varying refractive index across the core, the light ray bent smoothly and coverge rapidly at points along the cable.
- The light ray near the edge of the core take a longer path but travel faster since refractive index is less.
- All light paths tend to arrive at one point simultaneously.
- This results in less modal dispersion.
- It is not eliminated entirely, but the output pulse is not nearly as stretched as in multimode step-index cable.
- The output pulse is only slightly elongated.
- **As a result, this cable can be used at very high pulse rates and thus considerable amount of information can be carried on it.**

Advantages:

- This type of cable is also much wider in diameter with core size in the **50 to 100 μm range.**
  1. Therefore, it is easier to splice and interconnect.
  2. Cheaper and less-intense light sources may be used.

Application:

Used for very high pulse rates.

2.2.3 Single-mode Step Index Fiber

- A single mode fiber guides the light ray in straight line and has core of smaller diameter as compared to multi-mode fiber.
- In single mode (mono) light travels along single path.
- It is used for extremely high data rate applications like long distance telephone transmission. Here core has smaller diameter as shown in Fig. 2.8.

- Due to small core, the total number of paths through the core are minimized and modal dispersion is eliminated.
- Typical core size are 2 to 15 μm.
- With minimum refraction no stretching occurs.
- The output pulse has essentially the same duration as the input pulse.

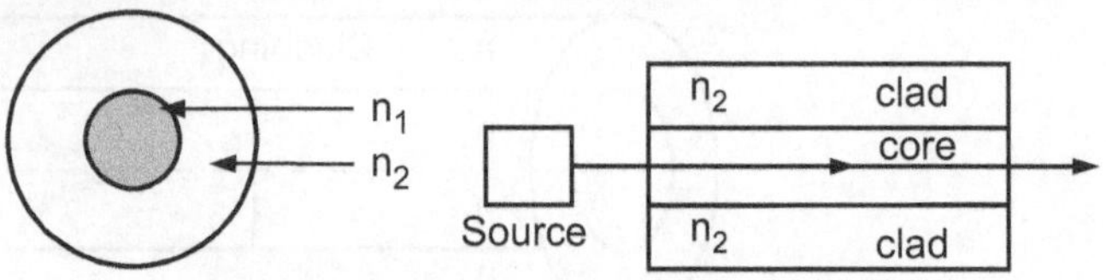

Fig. 2.8: Single Mode Step Index Fiber

Advantages:

The single mode step index fibers are best since the pulse repetition rate can be high and the maximum amount of information can be carried.

Disadvantages:

1. The main problem with this type of cable is because of very small size, it is difficult to make and is therefore very expensive.
2. Handling, splicing and making interconnections are more difficult.
3. Requires superintense light source such as laser.

Application:

For long distances single mode step index cable is used.

2.2.4 Comparison of Step Index and Graded Index Fiber

Sr. No.	Parameter	Step-Index Fiber	Graded-Index Fiber
1.	Data rate	Slow	High
2.	Coupling efficiency	High	Low
3.	Index variation	$\Delta = \frac{n_1 - n_2}{n_1}$	$\Delta = \frac{n_1^2 - n_2^2}{2n_1^2}$
4.	Ray path	By total internal reflection	Travels in oscillatory manner
5.	Numerical aperature	NA remain same.	Changes continuously with distance from fiber axis.
6.	Material	Plastic or glass	Only glass
7.	Attenuation	Less 0.34 dB/km	More 0.6-1 dB/km
8.	Light source	LED	LED, LASER
9.	Application	Subscriber local network communication	Local network and WAN

2.3 WAVE PROPAGATION

- Consider, the propagation of light in step-index fibers by using Maxwell's equations for electromagnetic waves.

2.3.1 Maxwell's Equations

- To analyse the optical waveguide, Maxwell's equations should have to be considered, that gives the relationship between electric and magnetic fields.
- For a non-conducting medium, without free charges and having no currents, these equations take the form in SI units.

$$\nabla \times E = \frac{-\partial B}{\partial t} \quad \ldots (1)$$

$$\nabla \times H = \frac{\partial D}{\partial t} \quad \ldots (2)$$

$$\nabla \cdot D = 0 \quad \ldots (3)$$

$$\nabla \cdot B = 0 \quad \ldots (4)$$

- where, **D** and **B** are corresponding to flux densities and **E** and **H** are electric and magnetic field vectors respectively.

$\therefore \quad D = \varepsilon E$

$\therefore \quad B = \mu H$

- The parameter, ε is the permittivity and μ is the permeability of the medium.

- Taking curl of equation (1) and making use of equation (2) yields

$$\nabla \times (\nabla \times E) = -\mu \frac{\partial}{\partial t} (\nabla \times H) = -\varepsilon\mu \frac{\partial^2 E}{\partial t^2} \qquad \ldots (5)$$

- Using vector identity,

$$\nabla \times (\nabla \times E) = \nabla (\nabla \cdot E) - \nabla^2 E$$

- Making use of equations (3), equation (5) becomes,

$$\nabla^2 E = \varepsilon\mu \frac{\partial^2 E}{\partial t^2} \qquad \ldots (6)$$

- By taking curl of equation (2), we have,

$$\nabla^2 H = \varepsilon\mu \frac{\partial^2 H}{\partial t^2} \qquad \ldots (7)$$

- Equations (6) and (7) are standard wave equations.

2.4 DISPERSION IN SINGLE MODE FIBERS

Dispersion of light is the distortion of light pulse and it is broadening of light pulse and merging of pulse into another pulse.

- Dispersion of the transmitted optical signal causes distortion for both digital and analog transmission along optical fibers.
- **Digital modulation used for optical fiber, then dispersion mechanisms within the fiber cause broadening of the transmitted light pulses as they travel along the channel.**
- From Fig. 2.9, it may be observed that each pulse broadens and overlaps with its neighbours, becoming indistinguishable at the receiver input. The effect is known as **Intersymbol Interference (ISI).**

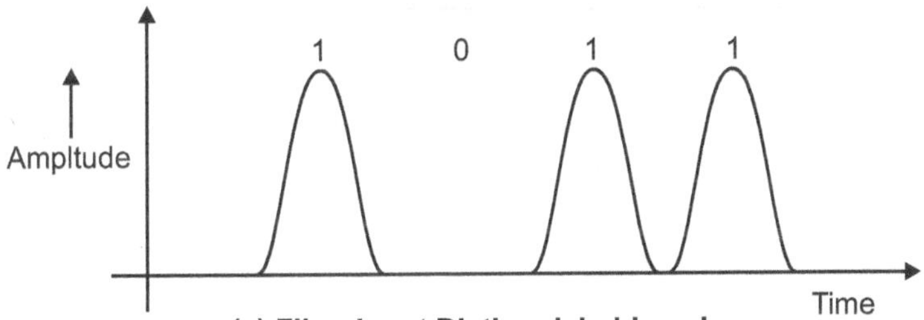

(a) Fiber input Distinguishable pulses

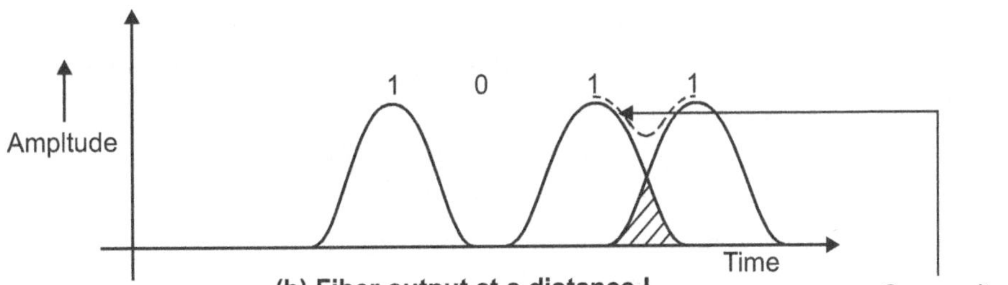

(b) Fiber output at a distance L_1

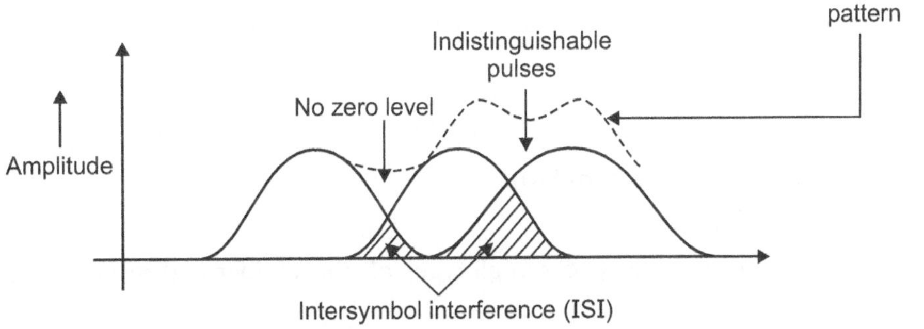

(c) Fiber output at a distance $L_2 > L_1$

Fig. 2.11: The Digital Bit Pattern 1011 of the Broadening of Light Pulses as they are transmitted along a Fiber

- Thus, number of errors may be encountered on the digital optical channel as the ISI becomes more pronounced.
- The error rate is also a function of the signal attenuation on the link and the subsequent signal to noise ratio (SNR) at the receiver.
- Signal dispersion alone limits the maximum possible bandwidth attainable with a particular optical fiber to the point where individual symbols can no longer be distinguished.
- For no overlapping of light pulses down on an optical fiber link the digital bit rate B_T must be less than the reciprocal of the broadened pulse duration (2τ). Hence,

$$B_T \leq \frac{1}{2\tau}$$

- This assumes that the pulse broadening due to the dispersion on the channel is τ, which dictates the input pulse duration that is also τ.
- Another more accurate estimate of the maximum bit rate for an optical channel with dispersion may be obtained by considering the light pulses at the output to have Gaussian shape with an r.m.s. width 6.
- **The effect of dispersion affects the maximum bit rate and hence the bandwidth.**
- A longer fibers results in more pulse broadening and hence leads to lower bandwidth.
- The dispersion also depends on the wavelength and bandwidth.
- For higher wavelength dispersion is less and hence bandwidth is more.
- **Types of Dispersion:**

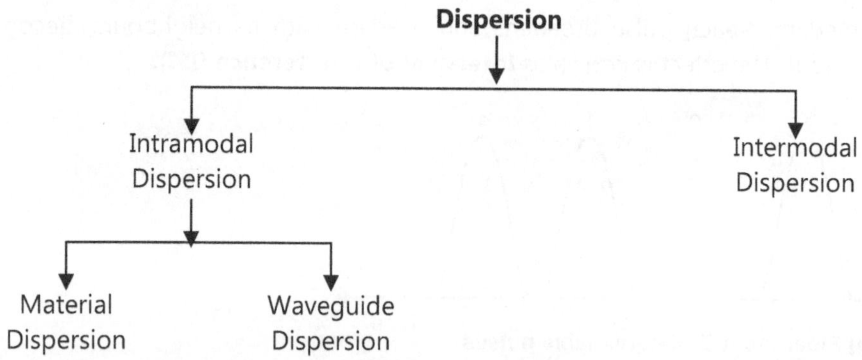

2.4.1 Group Velocity Dispersion (GVD)

- **GVD**, is the phenomenon that the group velocity of light in a transparent medium depends on the optical frequency or wavelength.

$$GVD = \frac{\partial}{\partial\omega}\frac{1}{v_g} = \frac{\partial}{\partial\omega}\left(\frac{\partial k}{\partial\omega}\right) = \frac{\partial^2 k}{\partial\omega^2}$$

- where, **k** is the frequency-dependent wavenumber. The group velocity dispersion is the **group delay dispersion** per unit length, denoted by $\mathbf{S^2/m}$.
- For optical fibers, the group velocity dispersion is usually defined as a derivative w.r.t. wavelength. This can be calculated as follows :

$$D_\lambda = \frac{\partial}{\partial_\lambda}\frac{1}{v_g} = -\frac{2\pi c}{\lambda^2}\cdot GVD = -\frac{2\pi c}{\lambda^2}\cdot\frac{\partial^2 k}{\partial\omega^2}$$

- This is usually specified with units of **PS/(nm km)**.

2.4.2 Intramodal Dispersion

- Intramodal or chromatic dispersion may occur in all types of optical fiber and results from the finite spectral linewidth of the optical source.
- Since optical sources do not emit just a single frequency but a band of frequencies then there may be propagation delay differences between the different spectral components of the transmitted signal.
- This causes broadening of each transmitted mode and hence intramodal dispersion.

- The delay differences may be caused by the dispersive properties of the waveguide material i.e. material dispersion and guidance effects within the fiber structure called waveguide dispersion.

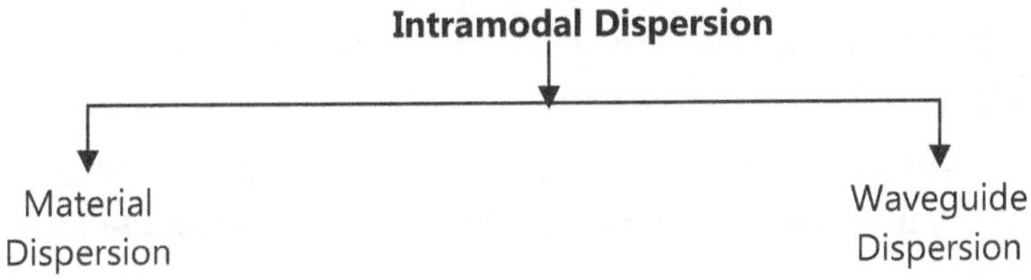

2.4.3 Material Dispersion

- Pulse broadening due to material dispersion results from the different group velocities of the various spectral components launched into the fiber from the optical source.
- It occurs when the phase velocity of a plane wave propagating in the dielectric medium varies non-linearly with wavelength and a material is said to exhibit material dispersion when the second differential of the refractive index with respect to wavelength is not zero.
- The pulse spread due to material dispersion may be obtained by considering the group delay τ_g in the optical fiber and is reciprocal of the group velocity v_g.
- Hence group delay.

  where,

$$\tau_g = \frac{d\beta}{d\omega} = \frac{1}{C}\left(n_1 - \lambda\frac{dn_1}{d\lambda}\right) \quad \dots (2.3)$$

$n_1 \rightarrow$ Refractive index of the core material

- Pulse delay τ_m due to material dispersion in a fiber of length is L is,

$$\tau_m = \frac{L}{C}\left(n_1 - \lambda\frac{dn_1}{d\lambda}\right) \quad \dots (2.4)$$

- Fig. 2.10 (a) shows the variation of the material dispersion parameter M with wavelength for pure silica.

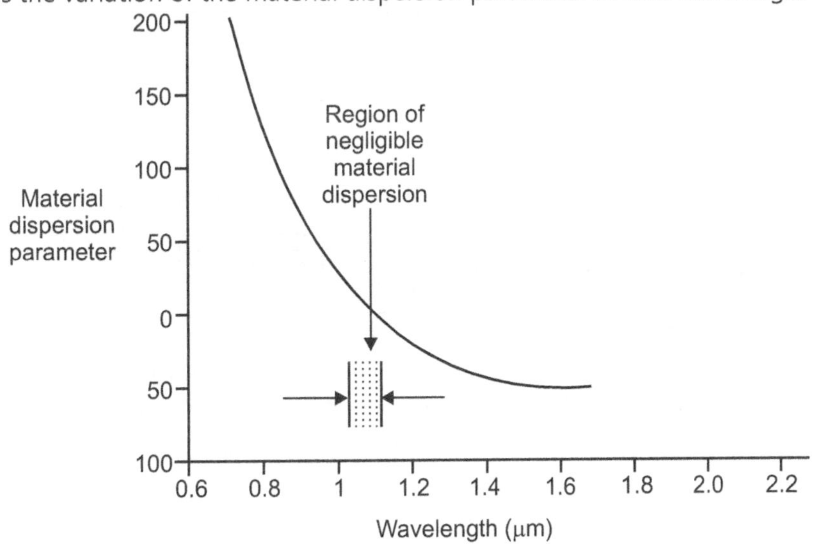

Fig. 2.10 (a): Material Dispersion Parameter for Silica

- It is observed that the material dispersion tends to zero in the longer wavelength region around 1.3 μm for pure silica.
- This provides an additional incentive for operation at longer wavelengths where the material dispersion may be minimized.
- Also, the use of an injection LASER with a narrow spectral width rather than LED as optical source leads to a substantial reduction in pulse broadening due to material dispersion, even in the shorter wavelength region.

2.4.4 Waveguide Dispersion

- The waveguide dispersion of fiber also creates intramodal dispersion.
- This results from the variation in group velocity with wavelength for a particular mode.

2.4.5 Intermodal Dispersion

- **Pulse broadening due to intermodal dispersion** (sometimes referred as simply modal or mode dispersion) **results from the propagation delay differences between modes within a multimode fiber.**
- As the different modes which constitute a pulse in a multimode fiber travels along the channel at different group velocities, the pulse width at the output is dependent on the transmission times of the slowest and fastest modes.
- This dispersion mechanism creates the fundamental differences in the overall dispersion for three types of fiber as shown in Fig. 2.10 (b).

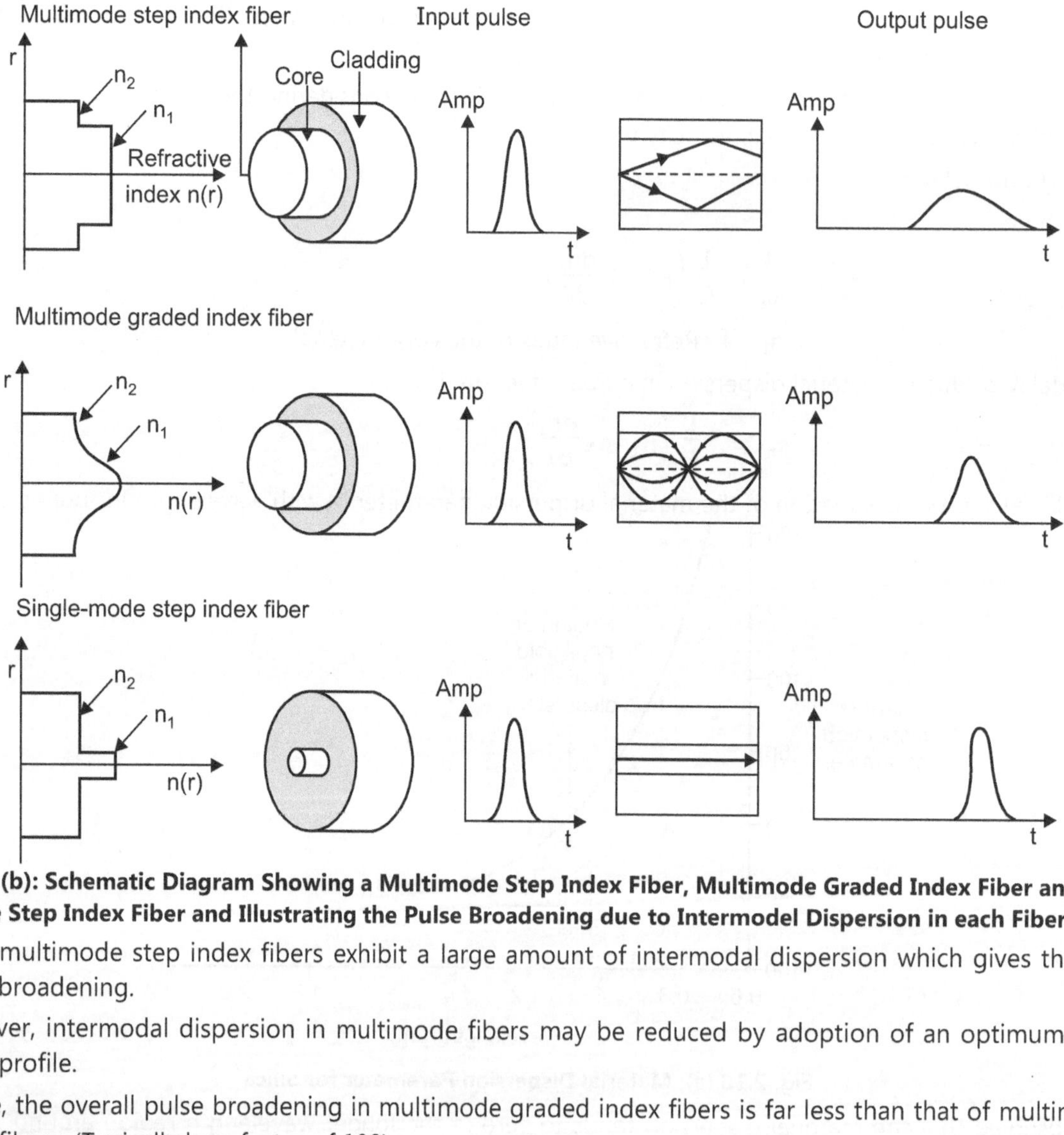

Fig. 2.10 (b): Schematic Diagram Showing a Multimode Step Index Fiber, Multimode Graded Index Fiber and Single-Mode Step Index Fiber and Illustrating the Pulse Broadening due to Intermodel Dispersion in each Fiber Type

- Thus, multimode step index fibers exhibit a large amount of intermodal dispersion which gives the greatest pulse broadening.
- However, intermodal dispersion in multimode fibers may be reduced by adoption of an optimum refractive index profile.
- Hence, the overall pulse broadening in multimode graded index fibers is far less than that of multimode step index fibers. (Typically by a factor of 100).
- **Thus, graded index fibers used with a multimode source give a tremendous bandwidth advantage or multimode step index fibers.**

2.4.6 Polarization-Mode Dispersion

- **Polarization-mode dispersion** results from the fact that light-signal energy at a given wavelength in a single-mode fiber actually occupies two orthogonal polarization states or modes.

- At the start of fiber, the two polarization states are aligned. However, since fiber material is not perfectly uniform, throughout its length, each polarization mode will encounter a slightly different refractive index.
- As a result, each mode will travel at a slightly different velocity. The result difference in propagation times between two orthogonal polarization modes will cause pulse spreading.

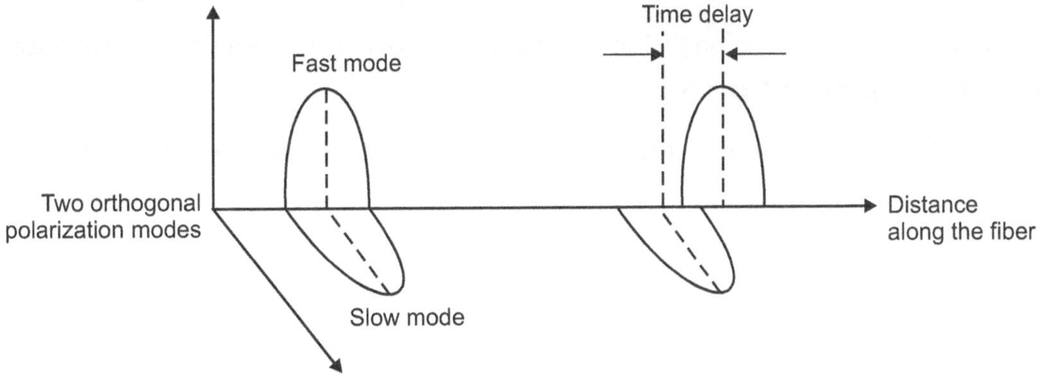

Fig. 2.11: Differences in Polarization Modes Propagation Times

2.5 DISPERSION INDUCED LIMITATIONS

- In dispersion induced limitation in the optical communication systems, when a fiber has a certain amount of dispersion, its immediate effect is to cause **pulse spreading**.

2.5.1 Basic Propagation Equation

- Each frequency component of optical field propagates in a single-mode fiber as,

$$\tilde{E}(r, \omega) = \hat{X} F(x, y)\, \tilde{B}(0, \omega) \exp(i\beta z) \quad \ldots (1)$$

where,

$\hat{X}$ = Polarization unit vector

$\tilde{B}(0, \omega)$ = Initial amplitude

β = Propagation constant

- Different spectral components of an optical pulse, propagate inside the fiber according to the simple relation,

$$\tilde{B}(z, \omega) = \tilde{B}(0, \omega) \exp(i\beta z) \quad \ldots (2)$$

The amplitude in the time domain is obtained by taking the inverse FT and is as follows:

$$B(z, t) = \frac{1}{2n} \int_{-\infty}^{\infty} \tilde{B}(z, \omega) \exp(-j\omega t)\, d\omega \quad \ldots (3)$$

- The initial spectral amplitude $\tilde{B}(0, \omega)$ is just the FT of the input amplitude B (0, t).
- Pulse broadening results from frequency dependence of β.

$$\beta(\omega) = \bar{\eta}(\omega) \frac{\omega}{c} \approx \beta_0 + \beta_1 (\Delta\omega) + \frac{\beta_2}{2} (\Delta\omega)^2 + \frac{\beta_3}{6} (\Delta\omega)^3 \quad \ldots (4)$$

- where,

$$\Delta\omega = \omega - \omega_o$$

$$\beta_m = \left(\frac{d^m\beta}{d\omega^m}\right)_{\omega=\omega_o}$$

- Substitute equation (2) and equation (4) in equation (3) and introduce a slowly varying amplitude A (z, t) of pulse envelope as

$$B(z, t) = A(z, t) \exp[i(\beta_0 z - \omega_0 t)] \quad \ldots (5)$$

- The amplitude A (z, t) is found as

$$A(z, t) = \frac{1}{2n} \int_{-\infty}^{\infty} j(\Delta\omega)\, \tilde{A}(0, \Delta\omega)$$

$$\exp\left[i\beta_1 z\Delta\omega + \frac{i}{2}\beta_2 z (\Delta\omega)^2 + \frac{i}{6}\beta_3 z (\Delta\omega)^3 - i(\Delta\omega)\, t\right] \quad \ldots (6)$$

- By calculating $\frac{\partial A}{\partial z}$ and exposing that $\Delta\omega$ is replaced by $i\left(\frac{\partial A}{\partial t}\right)$ in time domain, equation (6) becomes,

$$\frac{\partial A}{\partial z} + \beta_1 \frac{\partial A}{\partial t} + \frac{i\beta_2}{2}\frac{\partial^2 A}{\partial t^2} - \frac{\beta_3}{6}\frac{\partial^3 A}{\partial t^3} = 0 \quad \text{... (7)}$$

- This is the basic propagation equation that proves that pulse evolution inside a single-mode fiber.

2.5.2 Ghirped Gaussian Pulses

- Let us consider the propagation of Chirped Gaussian pulses inside the optical fibers by choosing the initial fields as,

$$A(0, t) = A_o \exp\left[-\frac{1 + ic}{2}\left(\frac{t}{T_0}\right)^2\right] \quad \text{... (8)}$$

- where, A_o = Peak amplitude
- The parameter T_0 represents the half-width at 1/e intensity point. It is related to the full-width at half-maximum (FWHM) of the pulse by the relation,

$$T_{FWHM} = 2(ln2)^{1/2}\, T_0 \approx 1.665\, T_0 \quad \text{... (9)}$$

- A pulse is said to be chirped, if its carrier frequency changes with time. The frequency change is related to the phase derivation, given by,

$$\delta\omega(t) = -\frac{\partial\phi}{\partial t} = \frac{c}{T_0^2}t \quad \text{... (10)}$$

- where, ϕ = Phase of A (0, t)
- Time-dependent frequency shift $\delta\omega$ is called **chirp**.

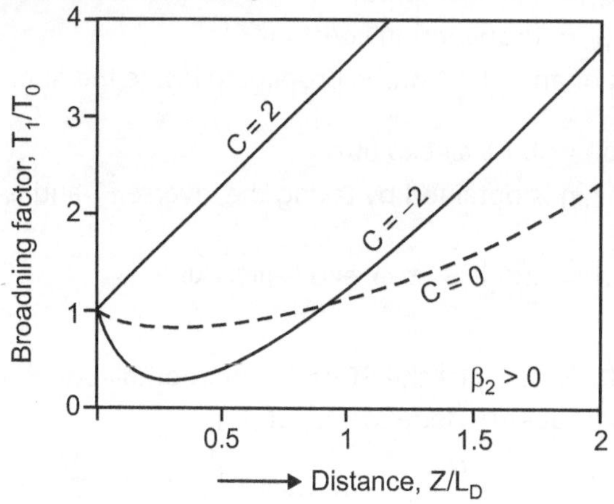

Fig. 2.12: Variation of Broadening Factor with Propagated Distance for a Chirped Gaussian Input Pulse

- By taking F.T. of equation (8), we have,

$$\tilde{A}(0, \omega) = A_0\left(\frac{2\pi T_0^2}{1 + ic}\right)^{1/2} \exp\left[-\frac{\omega^2 T_0^2}{2(1 + ic)}\right] \quad \text{... (11)}$$

- When the source spectrum is Gaussian with RMS spectral width σ_ω, the broadening factor is obtained as,

$$\frac{\sigma^2}{\sigma_0^2} = \left(1 + \frac{C\beta_2 L}{2\sigma_0^2}\right) + (1 + V_\omega^2)\left(\frac{\beta_2 L}{2\sigma_0^2}\right) + (1 + c^2 + V_\omega^2)^2\left(\frac{\beta_3 L}{4\sqrt{2}\,\sigma_0^3}\right) \quad \text{... (12)}$$

- where, $V_m = 2\,\sigma_\omega\,\sigma_0$. The equation (12) gives an expression for dispersion induced broadening of Gaussian input pulses.

2.5.3 Limitations of Bit Rate

- The limitations imposed on bit rate by fiber dispersion can be different, depending on the source spectral width.
- We can consider following two cases separately.

2.5.3.1 Optical Sources with Large Spectral Width

- This case corresponds to $V_\omega >> 1$ in equation (12). By setting c = 0 in equation (12), we have,

$$\sigma^2 = \sigma_0^2 (\beta_2 L \sigma_\omega)^2 \equiv \sigma_0^2 + (DL\sigma_\lambda)^2 \quad \text{... (13)}$$

- where, σ_λ is the RMS source spectral width in wavelength units. The output pulse width is given by,

$$\sigma = (\sigma_o^2 + \sigma_D^2)^{1/2} \quad \text{... (14)}$$

- where, $\sigma_D \equiv |D|\, L\, \sigma_\lambda$ provides a measure of dispersion induced broadening.
- When $\sigma_D >> \sigma_o$, the limitation on the bit rate given by,

$$BL\,|S|\,\sigma_\lambda^2 \leq 1/\sqrt{8} \quad \text{... (15)}$$

2.5.3.2 Optical Sources with a Small Spectral Width

- This case corresponds to $V_\omega << 1$ in equation (12). If we neglect the β_3 term and set c = 0, equation (12) can be given by,

$$\sigma^2 = \sigma_0^2 + \left(\frac{\beta_2 L}{2\sigma_0}\right)^2 \equiv \sigma_0^2 + \sigma_D^2 \quad \text{... (16)}$$

- If we compare with equation (14), gives a major difference between the two cases.

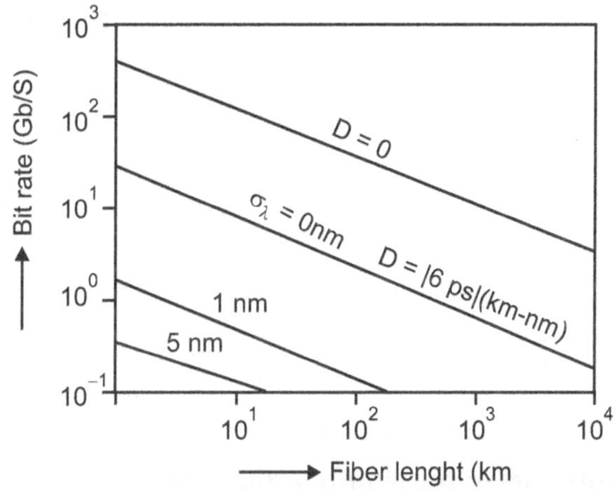

Fig. 2.13: Limiting Bit Rate of Single-Mode fibers as a Function of Fiber Length f_0 σ_λ = 0, 1 and 5 nm

2.5.4 Fiber Bandwidth

- The concept of fiber bandwidth originates from the theory of time-invariant linear systems.
- If the optical fiber can be treated as a linear system, its input and output power should be related by equation,

$$P_{out}(t) = \int_{-\infty}^{\infty} h(t - t')\, P_{in}(t')\, dt' \quad \text{... (17)}$$

- $P_{in}(t) = \delta(t)$, where $\delta(t)$ = Delta function. $P_{out}(t) = h(t)$. Hence, h(t) is called impulse response of linear system. Its T : T is

$$H(f) = \int_{-\infty}^{\infty} h(t) \exp(2\pi i f t)\, dt \quad \text{... (18)}$$

- The fiber BW f_3 dB corresponds to the frequency f = f_3 dB at which |H(f)| is reduced by a factor of 2 or 3 dB.

$$|H(f_3\, dB)/H(0)| = \frac{1}{2} \quad \text{... (19)}$$

- Note that, f_3 dB is the optical bandwidth of fiber as the optical power drops by 3 dB at this frequency compared with zero-frequency response.

2.6 LOSSES IN OPTICAL FIBER

- Here, we shall continue the discussion of optical fibers by answering two very important questions:
  1. What are the loss or signal attenuation mechanism in a fiber ?
  2. Why and to what degree do optical signals get distorted as they propagate along a fiber ?
- Signal attenuation known as fiber loss or signal loss is one of the most important properties of fiber optic, because it determines largely the maximum repeaterless separation between a transmitter and a receiver.
- Since the repeaters are expensive to fabricate, install and maintain, the degree of attenuation in a fiber has a large influence on system cost.
- Also equal importance is signal distortion.
- The distortion mechanisms in a fiber cause optical signal pulses to broaden as they travel along a fiber.
- If these pulses travel sufficiently long distance, they will eventually overlap with neighbouring pulses, thereby creating errors in the receiver output.
- The signal distortion mechanisms thus limits the information carrying capacity of a fiber.
- The predominant losses in optical fiber cable are:
  1. Absorption loss
  2. Scattering loss
  3. Dispersion loss
  4. Radiation loss
  5. Coupling loss

2.6.1 Attenuation

- The main specification of a fiber optic cable is its attenuation.
- **Attenuation means the loss of light energy as the light pulse travels from one end of cable to the other.**
- The light pulse of specific amplitude is applied to one end of the cable, but the light pulse output at the other end will be much lower in amplitude.
- The intensity of light at the output is low because of various losses in the cable.
- The reasons for the losses in light intensity over the length of the cable is due to light.
  - Absorption
  - Scattering and
  - Dispersion
- The amount of attenuation varies with the type of cable and its size.
- Glass has less attenuation than plastic.
- Wider cores have less attenuation than narrow cores.
- **Attenuation is directly proportional to the length of the cable.**
- It is obvious that the longer the distance the light has to travel, the greater the loss due to absorption, scattering and dispersion.
- Doubling the length of cable doubles the attenuation and so on.
- The attenuation of a fiber optic cable is expressed in logarithmic unit of decibel.

$$\because \quad dB = 10 \log_{10} \frac{P_o}{P_{in}} \quad \text{... (2.1)}$$

where, P_o = Power output

P_i = Power in

The decibel is used for comparing two power levels, may be defined for a particular optical wavelength as the ratio of the input (transmitted) optical power P_i into a fiber to the output (received) optical power P_o from the fiber.

Table 2.1 shows the percentage of output power for various decibel losses.

For example, 3 dB represents half power.

In other words, 3 dB loss means only 50% of the input appears at the output, the other 50% of the power is lost in the cable.

Table 2.1: Decibel Loss Table

Loss (dB)	Power Output (%)
1	79
2	63
3	50
4	40
5	31
6	25
7	20
8	14
9	12
10	10
20	1
30	0.1
40	0.01
50	0.001

- The higher the decibel figure, the greater the attenuation and loss.
- A 30 dB loss means only one-thousandth of the input power appears at the end.
- In fiber optic communications, the attenuation is usually expressed in decibels per unit length i.e. dB/km.

$$\therefore \quad \alpha_{dB} L = 10 \log_{10} \frac{P_i}{P_o} \qquad \ldots (2.2)$$

where,

α_{dB} = Signal attenuation per unit length in decibels

L = Fiber length

Example 2.1:

When the mean optical power launched into an 8 km length of the fiber is 120 μW, the mean optical power at the fiber output is 3 μW.

Determine:

(a) The overall signal attenuation or loss in decibels.

(b) The signal attenuation per km for the fiber.

(c) The overall signal attenuation for a 10 km optical link using the same fiber with splices at 1 km intervals, each giving an attenuation of 1 dB.

(d) The numerical input/output power ratio in (C).

Solution:

(a) Use equation 6.1, the overall attenuation in decibels through the fiber is,

$$\text{Signal attenuation} = 10 \log_{10} \frac{P_i}{P_o} = 10 \log_{10} \frac{120 \times 10^{-6}}{3 \times 10^{-6}} = \log_{10} 40 = \mathbf{16\ dB}$$

(b) The signal attenuation per km for the fiber may be obtained by dividing the results in (a) by the fiber length. Use equation 6.2.

$$\alpha_{dB} L = 16 \text{ dB}$$

$$\therefore \quad \alpha_{dB} = \frac{16}{8} = \mathbf{2\ dB/km}$$

(c) α_{dB} = 2 dB/km, the loss incurred along 10 km of fiber is,

$$\alpha_{dB} L = 2 \times 10 = \mathbf{20\ dB}$$

However, the link has nine splices (at 1 km intervals) each with an attenuation of 1 dB. Thus, loss due to splices is 9 dB. Hence, overall signal attenuation for link is,

$$\text{Signal attenuation} = 20 + 9 = 29 \text{ dB}$$

(d) To obtain a numerical value for input/output power ratio may be,

$$\frac{P_i}{P_o} = 10^{(dB/10)} = 10^{29/10} = \mathbf{794.3}$$

2.6.2 Absorption Loss

- **Material absorption is a loss mechanism related to the material composition and the fabrication process for the fiber, which results in dissipation of some of the transmitted optical power as heat in the waveguide.**
- The absorption of the light may be:

  (i) Intrinsic:

  Caused by the interaction with one or more of the components of glass.

  (ii) Extrinsic:

  Caused by impurities within the glass.

2.6.2.1 Intrinsic Absorption

- A pure silicate glass has small intrinsic absorption due to its basic material structure in the near infra-red region.
- However, it has two major intrinsic absorption mechanisms at optical wavelengths which leave a low intrinsic absorption window over the 0.8 to 1.7 μm wavelength range shown in Fig. 2.14.

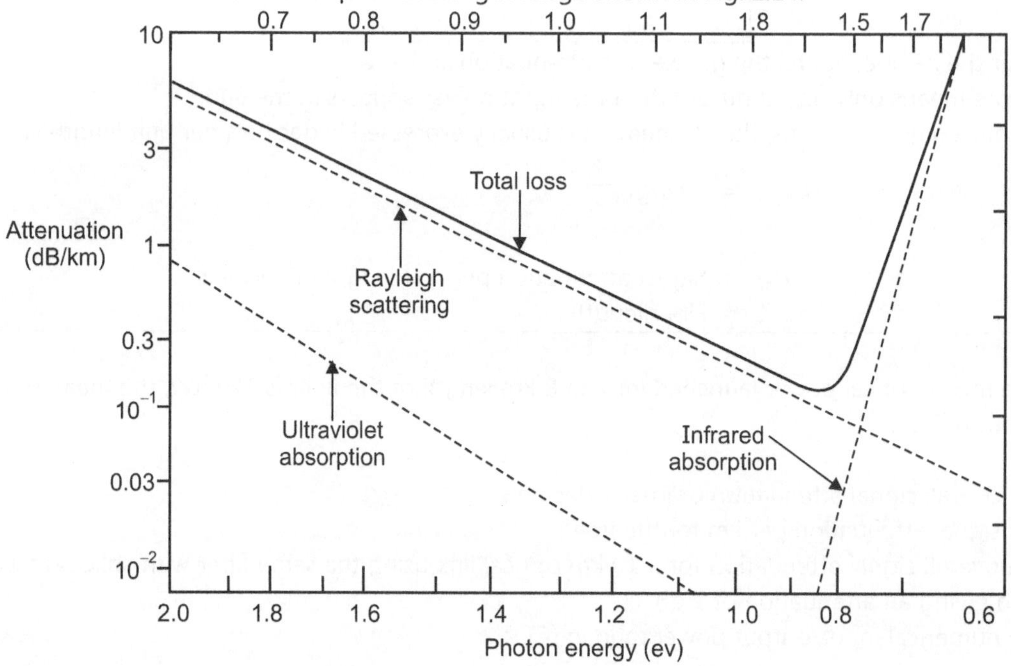

Fig. 2.14: The Attenuation Spectra for the Intrinsic Loss Mechanism in pure $GeO_2 - SiO_2$ Glass

- Fig. 2.14 shows a possible optical attenuation against wavelength characteristics for absolutely pure glass.
- It shows that there is a fundamental absorption edge, the peaks of which are centred in the ultraviolet wavelength region.
- Also in the infrared and far-infrared, at wavelength above 7 μm fundamentals of absorption bands from the interaction of photons with molecular variations within the glass occur.
- This give absorption peaks which again extends into the window region.
- The strong absorption band occurs due to oscillations of structural units such as Si-O (9.2 μm), P-O (8.1 μm), B-O (7.2 μm) and Ge-O (11 μm) within the glass.
- Hence, above 1.5 μm the tails of these largely for infrared absorption peaks tend to cause most of the pure glass losses.
- However, effects of both these processes may be minimized by suitable choice of both core and cladding compositions.
- In some non-oxide glasses as fluorides and chlorides, the infrared absorption peaks occur at much longer wavelengths which are well into the far-infrared (upto 50 μm), gives less attenuation to longer wavelength transmission compared with oxide glass.

2.6.2.2 Extrinsic Absorption

- **Extrinsic absorption result from the presence of impurities.**
- Some common metallic impurities found in glasses shown in Table 2.2 together with the absorption losses caused by one part in 10^9.

Table 2.2: Absorption losses caused by some of the more common metallic ion impurities in glasses

Ion Impurities	Peak Wavelength (nm)	One Part in 10^9 (dB/km)
C_r^{3+}	625	1.6
C_+^2	685	0.1
C_u^{2+}	850	1.1
Fe^{2+}	1100	0.68
Fe^{3+}	400	0.15
Ni^{2+}	650	0.1
Mn^{2+}	460	0.2
V^{4+}	725	2.7

- From Table 2.2, it may be noted that certain of these impurities, namely chromium and copper, in their worst valence state can cause attenuation in excess of 1 dB/km in the near-infrared region.
- Transition element contamination may be reduced to acceptable levels (i.e. one part in 10^{10}) by glass refining techniques such as vapour-pulse oxidation which eliminates the effects of these metallic impurities.
- However, another major **extrinsic loss mechanism** is caused by absorption due to water (i.e. OH ion) dissolved in the glass.
- These hydroxyl groups are bonded into the glass structure and have fundamental stretching vibrations which occur at wavelengths between 2.7 and 4.2 μm depending one group position in the glass network.
- The fundamental vibrations give rise to over tones appearing harmonically at 1.38, 0.95 and 0.75 μm shown in Fig. 2.15.

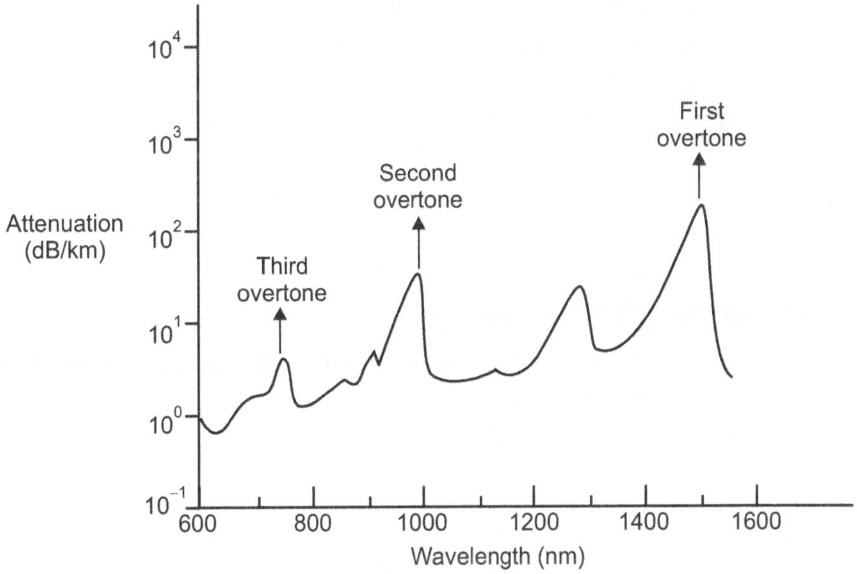

Fig. 2.15: The Absorption Spectrum for the Hydroxyl (OH) Group in Silica

- Fig. 2.15 shows the absorption spectrum for the hydroxyl group in silica.
- The combinations between the overtones and the fundamental SiO_2 vibration occur at 1.24, 1.13 and 0.88 μm completing the spectrum.
- It may also be observed that, the only significant absorption band in the region below a wavelength of 1 μm is the second overtone at 0.95 μm which causes attenuation at about 1 dB/km for one part per million of hydroxyl.
- At longer wavelengths the first overtone at 1.38 μm and its sideband at 1.24 μm are strong absorbers gives attenuation of 2 dB/km ppm and 4 dB/km ppm respectively.

2.7 SCATTERING LOSS

- **Scattering loss is one type of attenuation.**
- Scattering is the loss of optical energy due to imperfections in the fiber and from the basic structure of the fiber. It scatters the light in all directions shown in Fig. 2.16.

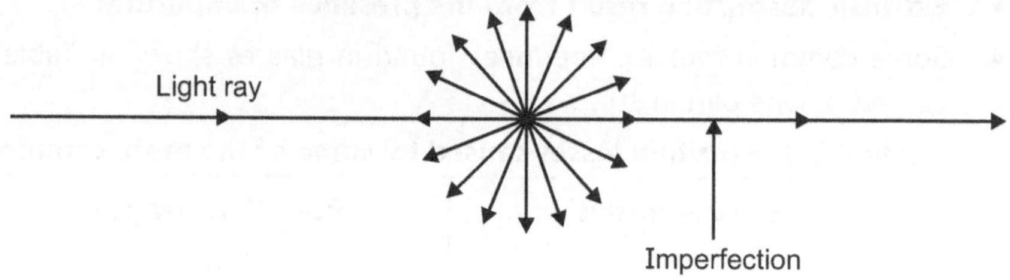

Fig. 2.16: Scattering of Light

Scattering is inversely proportional to the forth power of the wavelength $\left(\frac{1}{\lambda^4}\right)$.

Scattering Losses

- Linear Scattering Losses
  - Rayleigh Scattering
  - Mie Scattering
- Non-linear Scattering Losses
  - Stimulated Brillouin Scattering
  - Stimulated Raman Scattering

2.8 LINEAR SCATTERING LOSS

- Linear scattering mechanisms cause the transfer of some or all of the optical power contained within one propagating mode to be transferred linearly into different mode.
- This process tends to result in attenuation of the transmitted light as the transfer may be a leaky or radiation mode which does not continue to propagate within the fiber core, but is radiated from the fiber.
- Note that with all linear processes there is no change of frequency on scattering.
- Linear scattering categorized in two types:
  (i) Rayleigh scattering. (ii) Mie scattering.
- Both results from the non-ideal physical properties of the manufactured fiber which are difficult and in certain cases impossible to eradicate at present.

2.8.1 Rayleigh Scattering

- **Rayleigh scattering is the dominant intrinsic loss mechanism in the low absorption window between the ultraviolet and infrared absorption tails.**
- It results from inhomogeneities of a radom nature occurring on a small scale compared with the wavelength of light.
- These homogeneities manifest themselves as refractive index fluctuations, arise from density and compositional variations which are frozen into the glass lattice on cooling.
- These compositional variations may be reduced by improved fabrications, but the index fluctuations caused by the freezing in of density inhomogeneities are fundamental and cannot be avoided.
- **The subsequent scattering due to the density fluctuations, which is almost in all directions, produces an attenuation proportional to $\frac{1}{\lambda^4}$ following the Rayleigh scattering formula.**
- For a single component glass this is given as,

$$\gamma_R = \frac{8\pi^3}{3\lambda^4} n^8 \rho^2 B_c kT_f \qquad \dots (2.9)$$

where,

γ_R = Rayleigh scattering coefficient.

λ = Optical wavelength.

n = Refractive index of the medium.

ρ = Average photoelastic coefficient.

B_c = Isothermal compressibility at fictive temperature T_f.

k = Boltzmann's constant.

Here, **the fictive temperature is defined as the temperature at which the glass can reach a state of thermal equilibrium and is closely related to the anneal temperature.**

- Also, the Rayleight scattering coefficient is related to the transmission loss factor (i.e. transmissivity) of the fiber $\mathcal{L}$ following the reaction

$$\mathcal{L} = e \times \rho \, (-\gamma_R L) \qquad \dots (2.10)$$

where,

L = Length of the fiber

From equation (2.10), it is apparent that the fundamental component of Rayleigh scattering is strongly reduced by operating at the longest possible wavelength.

- The theoretical attenuation due to Rayleigh scattering in silica at wavelengths of 0.63, 1.00 and 1.30 μm respectively, these theoretical results are reasonable agreement with experimental work.
- The predictable attenuation due to Rayleigh scattering against wavelength is indicated by broken line on the attenuation characteristics shown in Fig. 2.17.

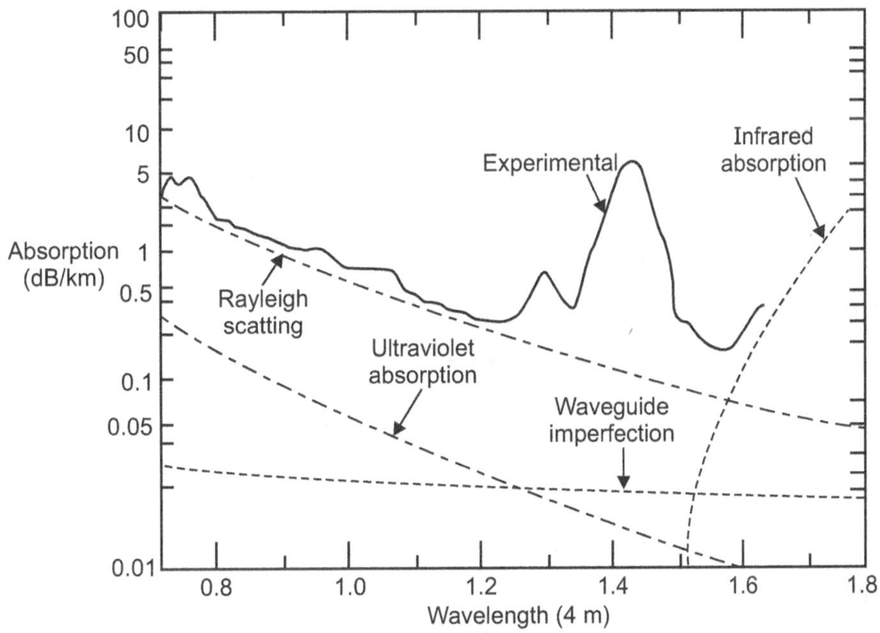

Fig. 2.17: Attenuation Spectrum for an Ultra-low-loss Single Mode Fiber with the Calculated Attenuation Spectra for some of the Loss Mechanism contributing to overall Fiber Attenuation

Example 2.2:

Silica has an estimated fictive temperature of 1400 K with an isothermal compressibility of 7×10^{-11} m^2/N. The refractive index and the photoleastic coefficient for silica and 1.46 and 0.286 respectively.

Determine the theoretical attenuation in decibels per km due to the fundamental Rayleigh scattering in silica at optical wavelengths of 0.63, 1.00 and 1.30 μm. Boltzman's constant is 1.381×10^{-2} J/K.

Solution:

Given:

$$T_f = 1400\ K$$
$$\beta_c = 7 \times 10^{-11}\ m^2/N$$
$$n = 1.46$$
$$\rho = 0.286$$
$$\lambda = 0.63,\ 1 \text{ and } 1.30\ \mu m$$
$$k = 1.381 \times 10^{-23}\ J/K$$

The Rayleigh scattering coefficient obtained from equation (6.3),

$$\gamma_R = \frac{8\pi^3 n^8 \rho^2 \beta_c\, k\, T_f}{3\lambda^4}$$

$$\therefore \quad \gamma_R = \frac{8\pi^3 \times (1.46)^8 \times (0.286)^2 \times 7 \times 10^{-11} \times 1.381 \times 10^{-3} \times 1400}{3 \times \lambda^4}$$

$$= \frac{248.15 \times 20.65 \times 0.082 \times 7 \times 10^{-11} \times 1.381 \times 10^{-23} \times 1400}{3\lambda^4} = \frac{1.895 \times 10^{-28}}{\lambda^4} / m$$

At a wavelength 0.63 μm

$$\gamma_R = \frac{1.895 \times 10^{-28}}{(0.63 \times 10^{-6})^4} = \frac{1.895 \times 10^{-28}}{0.158 \times 10^{-24}} = \mathbf{1.199 \times 10^{-3}\ /m}$$

The transmission loss factor one km of fiber obtained from equation (6.4).

$$\mathcal{L}_{km} = e \times \rho\,(-\gamma_R L) = e \times \rho\,(-1.199 \times 10^{-3} \times 10^3) = \mathbf{0.301}$$

The attenuation due to Rayleigh scattering in dB/m obtained from equation (6.1).

$$\text{Attenuation} = 10 \log_{10}\left(\frac{1}{\mathcal{L}_{km}}\right) = 10 \log_{10} 3.322 = \mathbf{5.2\ dB/km}$$

At a wavelength of 1.00 μm

$$\gamma_R = \frac{1.895 \times 10^{-28}}{10^{-24}} = \mathbf{1.895 \times 10^{-4}\ /m}$$

Using equation (6.4),

$$\mathcal{L}_{km} = e \times \rho\,(-1.895 \times 10^{-4} \times 10^3) = e \times \rho\,(-0.1895) = \mathbf{0.827}$$

From equation (6.1),

$$\text{Attenuation} = 10 \log_{10} 1.209 = 0.8\ dB/km$$

At a wavelength of 1.30 μm

$$\gamma_R = \frac{1.895 \times 10^{-28}}{2.856 \times 10^{-24}} = \mathbf{0.664 \times 10^{-4}}$$

Using equation (6.4),

$$\mathcal{L}_{km} = e \times \rho\,(-0.664 \times 10^{-4} \times 10^3) = \mathbf{0.936}$$

From equation (6.1),

$$\text{Attenuation} = 10 \log_{10} 1.069 = \mathbf{0.3\ dB/km}$$

2.8.2 Mie Scattering

- Linear scattering also occur at inhomogeneities which are compatible in size to the guided wavelength.
- These results from the non-perfect cylindrical structure of the waveguide and may be caused by fiber imperfections such as irregularities in the core-cladding interface, core-cladding refractive index differences along with fiber length, diameter fluctuations, strains and bubbles.
- When the scattering inhomogeneity size is greater than λ/10, the scattered intensity which has an angular dependence can be very large.
- **The scattering created by such inhomogeneities is mainly in forward direction, is called Mie scattering.**
- Depending on the fiber material, design and manufacture, Mie scattering can cause losses.

- The inhomogenities can be reduced by:
  (i) Removing imperfections due to the glass manufacturing process.
  (ii) Carefully controlled extrusion and coating of the fiber.
  (iii) Increasing the fiber guidance by increasing the relative refractive index difference.

  So that it is possible to reduce Mie scattering to insignificant levels.

2.9 NON-LINEAR SCATTERING

- Optical wavelength do not always behave as completely linear channels whose increase in output optical power is directly proportional to the input optical power.
- Usually at high optical power levels, several non-linear effects occur, which is in the case of scattering cause disproportionate attenuation.
- This non-linear scattering causes the optical power from one mode to the transferred in either the forward or backward direction to the same or other modes at different frequency.
- It depends critically on the optical power density within the fiber and hence become significant only above threshold power levels.
- There are **two important types of non-linear scattering** within optical fibers are:

  (i) Stimulated Brillouin and (ii) Raman scattering
- Both of these types usually observed only at high optical power densities in long single mode fibers.
- These scattering mechanisms give optical gain but with a shift in frequency, thus contributing to attenuation for light transmission at specific wavelength.
- Note that such non-linear phenomena can also be used to give optical amplification in the context of integrated optical techniques.

2.9.1 Stimulated Brillouin Scattering

- **Stimulated Brillouin Scattering (SBS) regarded as the modulation of light through thermal molecular vibrations within the fiber.**
- The scattered light appears as upper and lower sidebands which are separated from the incident light by the modulation frequency.
- The incident photon in this scattering process produces a photon of acoustic frequency as well as a scattered photon.
- This produces an optical frequency shift which varies with the scattering angle because the frequency of sound wave varies with acoustic wavelength.
- Frequency shift is a maximum in the backward direction reducing to zero in the forward direction making SBS and backward process.
- Brillouin scattering is significant only above a threshold power density, assuming that the polarization state of the transmitted light is not maintained i.e. the threshold power P_B is given by,

$$P_B = 4.4 \times 10^{-3} d^2 \lambda^2 \alpha_{dB} \gamma \text{ Watts} \quad \ldots (2.11)$$

where,

d = Fiber core diameter in μm.

λ = Operating wavelength in μm.

α_{dB} = Fiber attenuation in dB/km.

γ = Source bandwidth in GHz.

Equation (2.11) allows the determination of the threshold optical power which must be launched into a single-mode optical fiber before SBS occurs.

2.9.2 Stimulated Raman Scattering

- **Stimulated Raman Scattering (SRS)** is similar to SBS except that a high frequency optical photon rather than an acoustic photon is generated in the scattering process.
- SRS can occur in both forward and backward directions in the optical fiber and may have an optical power threshold of upto three orders of magnitude higher than the Brillouin threshold in a particular fiber.

- Using some criteria as those specified in SBS threshold given in equation (2.11), it may be shown that the threshold optical power for SRS in a long single mode fiber is given by,

$$P_R = 5.9 \times 10^{-2} d^2\lambda\, \alpha_{dB} \text{ Watts} \quad \text{... (2.12)}$$

- SBS and SRS are not usually observed in multimode fibers because of their large core diameter make the threshold optical power levels extremely high.
- Note that the threshold optical powers for both these scattering mechanisms may be increased by suitable adjustment of the other parameters in equations (2.11) and (2.12).
- Operation at the longest possible wavelength is advantageous.

Example 2.3:

A long single-mode optical fiber has an attenuation of 0.5 dB/km when operating at a wavelength of 1.3 μm. The fiber core diameter is 6 μm and the laser source bandwidth is 600 MHz. Compare the threshold optical powers for Stimulated Brillouin and Raman Scattering within the fibers at the specified wavelength.

Solution:

Given:

$$d = 6 \ \mu m$$
$$\lambda = 1.3 \ \mu m$$
$$\alpha_{dB} = 0.5 \text{ dB/km}$$
$$\gamma = 600 \text{ MHz}$$

The threshold optical power for SBS is,

$$P_B = 4.4 \times 10^{-3} d^2\lambda^2 \alpha_{dB}\, \nu = 4.4 \times 10^{-3} \times 6^3 \times 1.3^2 \times 0.5 \times 0.6 = \mathbf{80.3\ mV}$$

The threshold optical power for SRS is,

$$P_R = 5.9 \times 10^{2} d^2 \lambda\, \alpha_{dB} = 5.9 \times 10^{-2} \times 6^2 \times 1.3 \times 0.5 = \mathbf{1.38\ W}$$

2.10 WAVEGUIDE IMPERFECTIONS

- In practice, imperfections at the core-cladding interface can lead to additional losses which contribute to the net fiber loss.
- The physical process behind such losses is Mie scattering, occurring because of index in homogeneities on a scale longer than the optical wavelength.
- The major source of fiber loss, particularly in cable form, is related to the random axial distortions such losses are referred as microbending losses.
- Microbends cause on increase in fiber loss for both multimode and single mode fibers and results in excessively large loss.
- Many other sources of optical loss exists in a fiber cable. These are related to splice and connectors used in forming the fiber link and are often treated as a part of cable loss. Microbending losses can also be included in the total cable loss.

2.11 NON-LINEAR PHASE MODULATION

- The non-linear part of refractive index, significantlty affects modern lightwave systems due to long fiber lengths.
- It leads to the phenomena of self-phase modulation (SPM) and cross-phase modulation (XPM).
- Both SPM and XPM produces a phase alteration of pulse, which leads to the change in its spectrum.
- SPM determines a broadening of pulse spectrum due to its own intensity.
- SPM is a non-linear phenomenon affecting the phase of propagating wave and does not involve other waves.

2.12 FOUR WAVE MIXING

- Four wave mixing is a non-linear effect arising from a third-order optical non-linearity, denoted by $x^{(3)}$ coefficient. It can occur if atleast two different frequency components propagate together in a non-linear medium such as optical fiber.
- Assuming just two input frequency components, v_1 and v_2 with $v_2 > v_1$), a refractive index modulation at the different frequency occurs, which creates two additional frequency components as shown in Fig. 2.18.
- In effect, two new frequency components are generated $v_3 = v_1 - (v_2 - v_1) = 2v_1 - v_2$ and $v_4 = v_2 + (v_2 - v_1) = 2v_1 - v_1$.

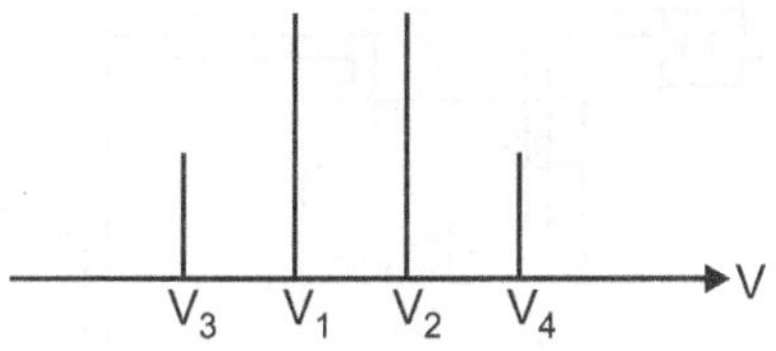

Fig. 2.18: Generation of New Frequency Components

- Four wave mixing in fibers is related to self-phase modulation and cross-phase modulation.

2.13 FIBER MANUFACTURING

- Because of rapidly increasing demands for telephone communication throughout the world, multiconductor copper cables have become not only very expensive but also insufficient way to meet these information requirements.
- The frequency limitations in the copper conductor system approximately 1 MHz.
- The optical fiber with its low weight and high frequency characteristics approximately 40 GHz and its imperviousness to interference from electromagnetic radiation, have become the choice for all heavy demand long-line telephone communication systems.

Why to use Optical Fibers ?

Reasons are:

1. The light weight and non-corrosiveness of fiber make it very practical for aircraft and automotive applications.
2. A single fiber can handle as many voice channels as a 1500 pair cable can.
3. The spacing of repeaters from 35 to 80 km for fibers as opposed to from 1 to $1\frac{1}{2}$ km for wire is great advantage.
4. Fiber is immune to interference from lightning, cross-talk and electromagnetic radiation.
   - In order to plan the use of optical fibers in a variety of line communication applications, it is necessary to consider the various optical fibers currently available.
   - The performance characteristics of the various fiber types vary considerably depending upon the materials used in the fabrication process and the preparation technique involved.
   - In particular, high performance silica-based fibers for operation in three major wavelength regions 0.8 to 0.9, 1.3 and 1.55 μm are now widely commercially available.
   - Complex refractive index profile single-mode fibers including dispersion modified fibers and polarization maintaining fibers are also commercially available and in the former case are starting to find system application within communications.
   - Another relatively new area of commercial fiber development is concerned with mid-infrared wavelength range 2 to 5 μm, often employed heavy metal fluoride glass technology.
   - However, the fiber products that exists for this wavelength region tend to multimode with relatively high losses and hence at present are only appropriate for special applications.
   - Several methods are used to make perform. The three commonly used methods are:
     - Modified Chemical-Vapour Deposition (MCVD).
     - Outside-Vapour Deposition (OVD).
     - Vapor-Axial Deposition (VAD).

(a) MCVD:

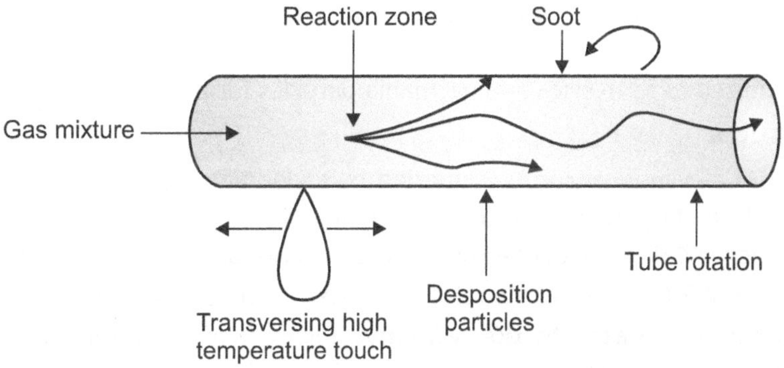

Fig. 2.19

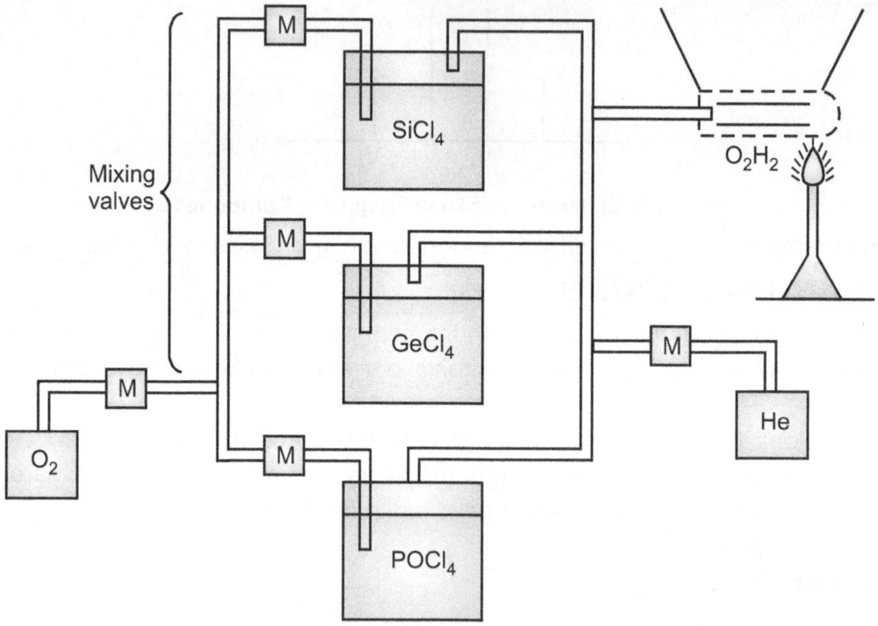

Fig. 2.20: MCVD Process commonly used for Fiber Fabrication

Explanation:

- First, a cylindrical perform is made by depositing layers of specially formulated silicon dioxide on the inside surface of hollow rod.
- The layers are deposited by applying a stream of pure oxygen to the substrate rod various chemical vapours, such as $SiCl_4$, $GeCl_4$ and $PoCl_3$ are added to the stream of oxygen.
- As the oxygen, contacts the hot surface of a rod, a flame underneath the rod keeps the walls of rod very hot. Silicon dioxide of high purity is formed.
- After sufficient layers are built up, the tube is collapsed into a solid glass rod referred to as perform.

(b) OVD : (Outside Vapour Deposition):

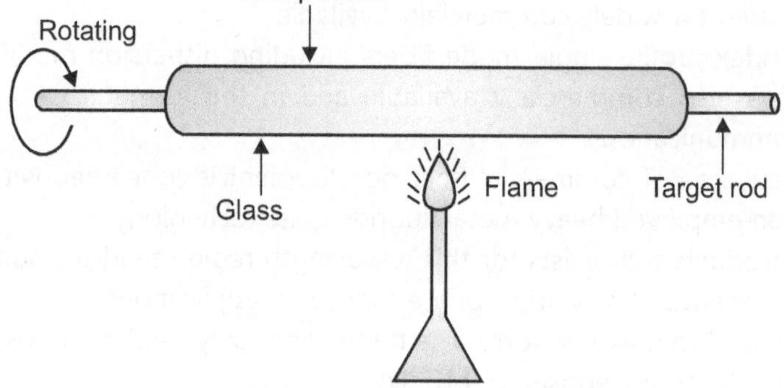

Fig. 2.21: Outside Vapour Deposition

Explanation:

- Here an inert rod is layered with core and cladding glass, deposits built up on the outside.
- Once enough layers are in place, the rod is removed and the layers and consolidate into a solid perform, which can be drawn into fiber.
- $SiCl_4$ and $GeCl_4$ are oxidized to form silica and germania particles for the deposition.

2.13.1 Design Issues

- A step-index consists of a cylindrical core surrounded by a cladding layer whose index is slightly lower than the core. Both core and cladding use silica as the base material.
- The differences in refractive indices is realized by doping the core or cladding or both.
- Dopants, such as GeO_2 and P_2O_5 increase the refractive index of silica and are suitable for core.
- **The major design issue are related to the refractive index profile, the amount of dopants and core, cladding dimensions.**

2.13.2 Fabrication Methods

- There are two main steps in the process of converting raw materials into optical fiber ready to be shipped.
  (1) Manufacturing of the pulse glass perform. (2) Drawing of perform.
- The first step in manufacturing glass optical fibers is to make a solid glass rod, known as a perform. $SiCl_4$ and $GeCl_4$ are converted into glass during perform manufacturing.
- The basic chemical reaction of manufacturing optical glass is,

$$SiCl_4 \text{ (gas)} + O_2 > SiO_2 \text{ (solid)} + 2Cl_2 \quad \text{... (a)}$$

$$GeCl_4 \text{ (gas)} + O_2 > GeO_2 \text{ (solid)} + 2Cl_2 \quad \text{... (b)}$$

2.14 FIBER CONNECTOR

- Fiber connectors are special mechanical assemblies that allow fiber optic cable to be connected to one another.
- A variety of connectors provide a convenient way to splice cables and attach them to transmitters, receivers and repeaters.

Properties:

1. Connectors force better alignment of the cables.
2. Two ends of cables must be aligned with precision so that excessive light is not lost. Otherwise, a splice or connection will introduce excessive attenuation.
3. Fiber optic connectors are the optical equivalent of electrical plugs and sockets.
4. They are mechanical assemblies that hold the ends of the cable and cause them to be accurately aligned with another cable.
5. Most of them either snap together or have threads which allow the two pieces to be screwed together.
6. Optical connectors corrects the problems of misalignment.

2.14.1 Ferrule Connector

- A typical Ferrule connector is shown in Fig. 2.22 (a).
- One end of the connector called Ferrule holds the fiber securely in place.

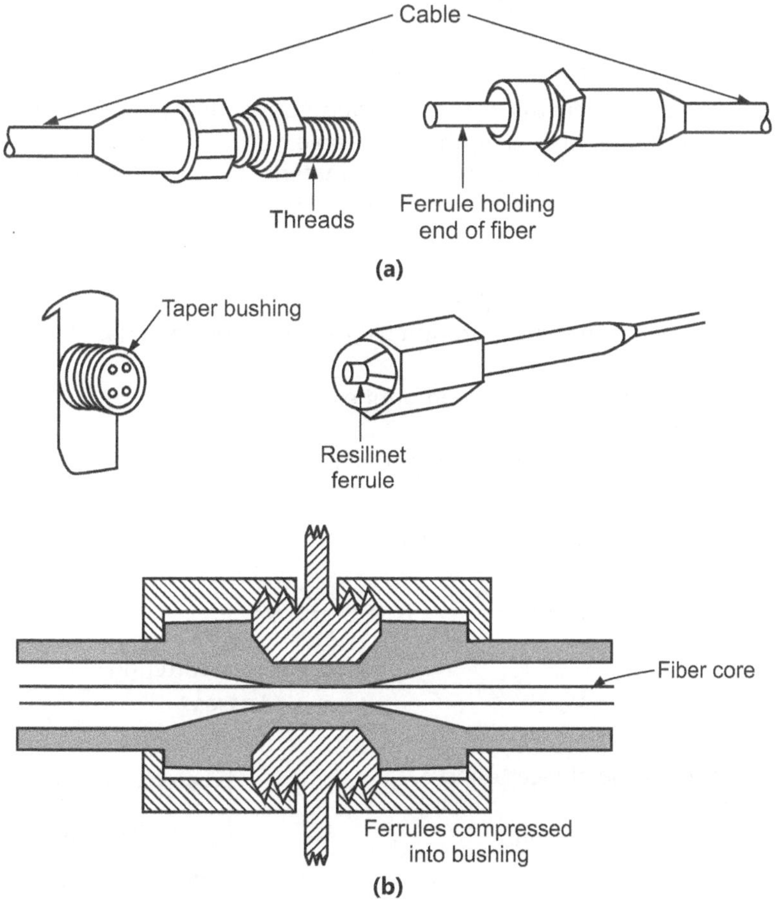

Fig. 2.22: Details of Fiber-Optic Ferrule Connector

- A matching fitting holds the other fiber securely in place.
- When the two are screwed together, the ends of the fiber touch, providing a low-loss coupling.
- Fig. 2.22 (b) shows how the connector aligns the fibers.
- Connectors are normally used at the end of the cable applied to the light source or at the end of the cable connected to the photo detectors.
- Connectors are also used at the repeater units where the light is picked-up, converted into an electrical pulse, amplified and reshaped and then used to create a new pulse to continue the transmission over a long line.

Important Points

- A single fiber can handle as many voice channels as a 1500 pair cable can.
- Fiber is immune to interference from lightning, cross-talk and electromagnetic radiation.
- The characteristics of light transmission through a glass fiber depends on many factors such as:
  1. The composition of the fiber.
  2. The amount and type of light introduced into the fiber.
  3. The diameter and length of the fiber.
- There are two major indices:
  1. Step index.
  2. Graded index.
- The **Refractive Index profile** describes the relation between the indices of core and cladding.
- **Mode:** Mode is a mathematical and physical concept desiring the propagation of electromagnetic wave through optical fiber.
- The fiber which has uniform distribution of refractive index n1 throughout its core is called as Step Index Fiber.
- **Multimode Propagation** in Step Index consists of many light rays.
- Multimode-step-index is widely used for short to medium distances at relatively low pulse frequencies.
- The core index is highest at center and decreases until it matches that of cladding.
- As a result, this cable can be used at very high pulse rates and thus considerable amount of information can be carried on it.
- A single mode fiber guides the light ray in straight line and has core of smaller diameter as compared to multi-mode fiber.
- The predominant losses in optical fiber cable are:
  1. Adsorption loss
  2. Scattering loss
  3. Dispersion loss
  4. Radiation loss
  5. Coupling loss
- Attenuation means the loss of light energy as the light pulse travels from one end of cable to the other.
- Attenuation is directly proportional to the length of the cable.
- Material absorption is a loss mechanism related to the material composition and the fabrication process for the fiber, which results in dissipation of some of the transmitted optical power as heat in the waveguide.
- The absorption of the light may be:
  (i) Intrinsic: Caused by the interaction with one or more of the components of glass.
  (ii) Extrinsic: Caused by impurities within the glass.
- A pure silicate glass has small intrinsic absorption due to its basic material structure in the near infra-red region.
- Extrinsic absorption result from the presence of impurities.
- Scattering loss is one type of attenuation.
- Scattering is the loss of optical energy due to imperfections in the fiber and from the basic structure of the fiber.
- 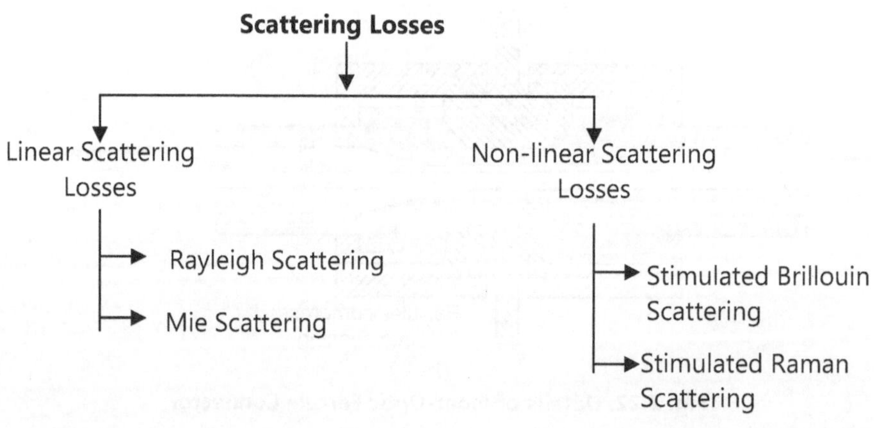

- Linear scattering mechanisms cause the transfer of some or all of the optical power contained within one propagating mode to be transferred linearly into different mode.
- Linear scattering categorized in two types:
  (i) Rayleigh scattering. (ii) Mie scattering.
- Rayleigh scattering is the dominant intrinsic loss mechanism in the low absorption window between the ultraviolet and infrared absorption tails.
- The subsequent scattering due to the density fluctuations, which is almost in all directions, produces an attenuation proportional to $\frac{1}{\lambda^4}$ following the Rayleigh scattering formula.
- The scattering created by such inhomogeneities is mainly in forward direction, is called Mie scattering.
- Optical wavelength do not always behave as completely linear channels whose increase in output optical power is directly proportional to the input optical power.
- There are two important types of non-linear scattering within optical fibers are:
  (i) Stimulated Brillouin and (ii) Raman scattering
- Stimulated Brillouin Scattering (SBS) regarded as the modulation of light through thermal molecular vibrations within the fiber.
- **Stimulated Raman Scattering (SRS)** is similar to SBS except that a high frequency optical photon rather than an acoustic photon is generated in the scattering process.
- Dispersion of light is the distortion of light pulse and it is broadening of light pulse and merging of pulse into another pulse.
- Digital modulation used for optical fiber, then dispersion mechanisms within the fiber cause broadening of the transmitted light pulses as they travel along the channel.
- 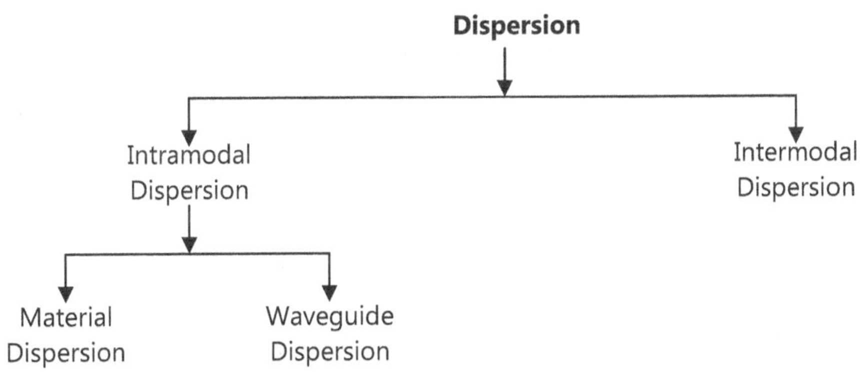

- Intramodal or chromatic dispersion may occur in all types of optical fiber and results from the finite spectral linewidth of the optical source.
- 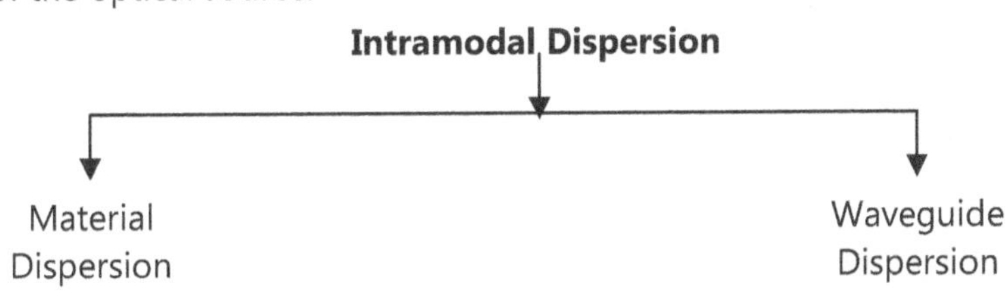

- Pulse broadening due to material dispersion results from the different group velocities of the various spectral components launched into the fiber from the optical source.
- The waveguide dispersion of fiber also creates intramodal dispersion.
- This results from the variation in group velocity with wavelength for a particular mode.
- **Pulse broadening due to intermodal dispersion** (sometimes referred as simply modal or mode dispersion) **results from the propagation delay differences between modes within a multimode fiber.**
- Radiation losses are caused mainly by small bends and links in the fiber.
- Coupling losses are caused by imperfect physical connections.

- Junction losses are most often caused by one of the following alignment problems.
  1. Lateral misalignment
  2. Gap misalignment
  3. Angular misalignment and
  4. Imperfect surface finishes
- **The basic transmitter is known as source.**
- **Conventional light sources such as incandescent lamps cannot be used in fiber optic systems** because they are simply too slow.
- An LED is a p-n junction diode usually made from a semiconductor material such as aluminium-gallium-arsenide (AlGaAs) or gallium-arsenide-phosphide (GaAsP).
- LEDs emit light by spontaneous emission, light emitted as a result of the recombination of electrons and holes.
- **Principle:** Laser action is the result of three process absorption of energy packets (photons), spontaneous emission, and stimulated emission. (These processes are represented by the simple two-energy-level diagrams).
- Predominant types of receiver used in communication system rely on principle of ionization of semiconductor.
- PIN diode can convert one photon to one electron of current. Hence, amplifier is needed with receiver.

Photo diodes

PIN Photo diode | Avalanche | Photo Diode

- **Principle:** Principle of PIN diodes is reverse that of LED. That is, light is absorbed at p-n junction rather than emitted.
- APDs amplify the signal during the detection process. They use a similar principle to that of photomultiplier tubes used in nuclear detection.

Practice Questions

1. Explain multimode step index fiber.
2. Compare multimode and single mode step index fiber.
3. Explain single mode step index fiber.
4. Explain stimulated Brillouin scattering.
5. Describe stimulated Raman scattering.
6. Explain Rayleigh scattering.
7. Explain group velocity and waveguide dispersion.
8. Give basic propagation equation.
9. Describe polarization mode dispersion.
10. Distribute fiber losses in detail.
11. Give attenuation coefficient in fiber loss.
12. Give expression for chirped Gaussian pulses.
13. Explain the steps in fiber manufacturing.
14. Give design issues in detail.
15. Explain fiber connectors.
16. Explain Dispersion-induced limitations.

***

Chapter 3

OPTICAL TRANSMITTERS BASIC CONCEPTS

3.1 INTRODUCTION

- The role of optical transmitter is to convert an electrical input signal into the corresponding optical signal and then launch it into the optical fiber saving as a communication channel.
- The major component of optical transmitter is an optical source.
- Fiber optical communication systems often use semiconductor optical sources such as light emitting diodes (LEDs) and semiconductor lasers because of several inherent advantages offered by them.
- In this chapter, LEDs and semiconductor lasers with their applications in lightwave system are explained.

3.2 LASER (EMISSION AND ABSORPTION RATES) (LIGHT AMPLIFICATION BY STIMULATED EMISSION OF RADIATION)

- The laser is a device which amplifies the light, hence the LASER is an acronym for light amplification by stimulated emission of radiation.
- The operation of the device may be described by the formation of an electromagnetic standing wave within a cavity (optical resonator) which provides an output of monochromatic highly coherent radiation.

Principle:

- **Laser action is the result of three process absorption of energy packets (photons), spontaneous emission, and stimulated emission.**
- **These processes are represented by the simple two-energy-level diagrams.**

  where, E1 is the lower state energy level.

  E2 is the higher state energy level.

- Quantum theory states that any atom exists only in certain discrete energy state, absorption or emission of light causes them to make a transition from one state to another.
- The frequency of the absorbed or emitted radiation is related to the difference in energy E between the two states.

  If E1 is lower state energy level

  and E2 is higher state energy level

  $E = E2 - E1 = h \cdot f$

  where, $h = 6.626 \times 10^{-34}$ Js (Planck's constant).

- An atom is initially in the lower energy state, when photon with energy E2 – E1 is incident on the atom it will be excited into the higher energy state E2 through the absorption of the photon.

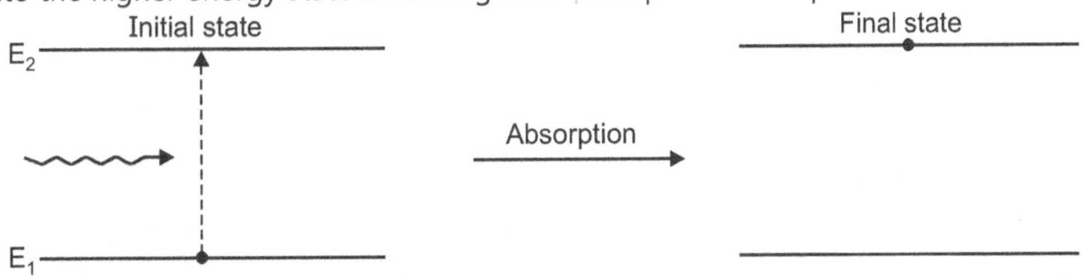

Fig. 3.1: Absorption

- When the atom is initially in the higher energy state E2, it can make a transition to the lower energy state E1 providing the emission of a photon at a frequency corresponding to $E = h \cdot f$.

- The emission process can occur in two ways:
  - (a) By spontaneous emission in which the atom returns to the lower energy state in random manner.
  - (b) By stimulated emission when a photon having equal energy to the difference between the two states (E2 – E1) interacts with the atom causing it to the lower state the creation of the second photon.

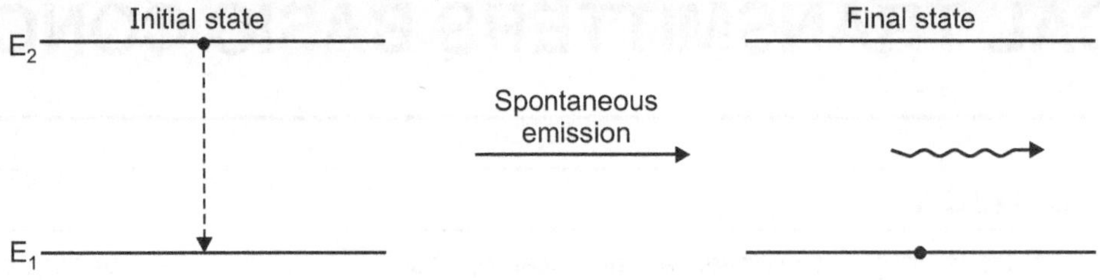

Fig. 3.2 (a): Spontaneous Emission

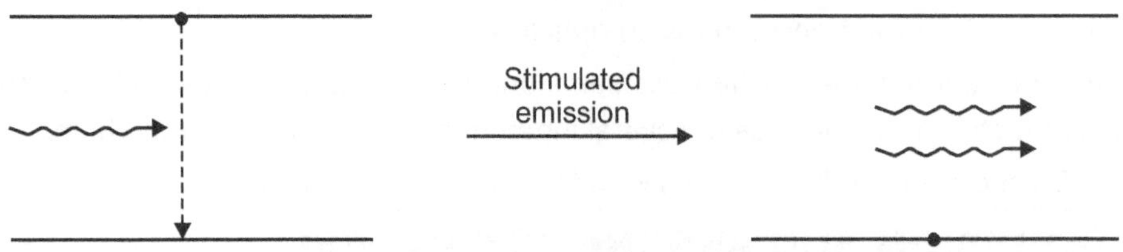

Fig. 3.2 (b): Stimulated Emission

- Spontaneous emission gives incoherent radiation.
- Stimulated emission gives coherent radiation.
- Since the light associated with emitted photon is of same frequency of incident photon, and in same phase with same polarization.
- It means that when an atom is stimulated to emit light energy by an incident wave, the liberated energy can add to the wave in constructive manner.
- The emitted light is bounced back and forth internally between two reflecting surfaces.
- The bouncing back and forth of light wave causes their intensity to reinforce and build-up.
- The result is a high brilliance, single frequency light beam providing amplification.

Construction:

- LASERs are constructed for many different materials, including gases, liquids and solids.
- The construction of LASER is similar to LED except that the ends are highly polished.
- The mirror like ends trap the photons in the active region and as they reflect back and forth, stimulate free electrons to recombine with holes at higher-than normal energy level. This is called lasing.

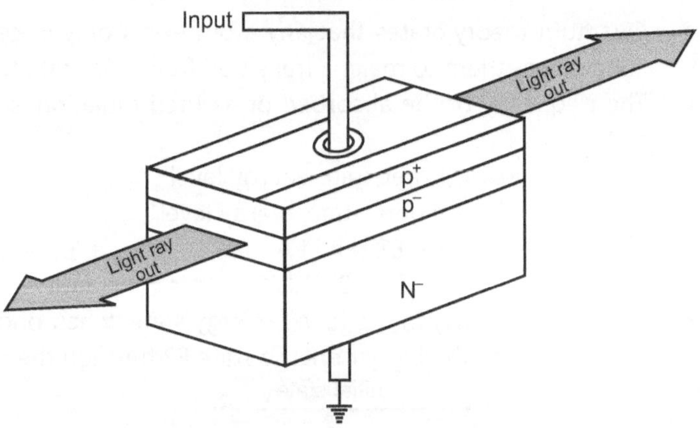

Fig. 3.3: LASER Diode Construction

Working:

- As current passes through a forward biased P-N junction diode, light is emitted by spontaneous emission at a frequency determined by the energy gap of the semiconductor material.
- When a particular current level is reached the number of minority carrier and photons produced on either side of the P-N junction reaches a level where they begin to collide with already excited minority carriers.
- This causes an increase in the ionization energy level and makes the carriers unstable.
- When this happens, a typical carrier recombine with an opposite type of carrier at an energy level that is above its normal before collision value.

- In the process, two photons are created, one is stimulated by another.
- Essentially, a gain in the number of photons is realized.
- For this to happen, a large forward current that can provide many carriers (holes and electrons) is required.

Advantages over LED:

1. Developing light power upto several watts.
2. More powerful than LEDs and therefore capable of transmitting much lower distances.
3. Ability to turn OFF and ON at a faster rate.
4. High speed laser diodes capable of gigabit per second digital data rates.

3.3 p-n JUNCTION

- The heart of a semiconductor optical source is the p-n junction, which is formed by bringing a p-type and n-type semiconductor into contact.
- Semiconductor is made by n-type or p-type by doping it with impurities, whose atoms have an excess valence electron.
- In case of n-type semiconductor, the excess electron occupy the conduction bond states, normally empty in intrinsic semiconductors.
- Fig. 3.4 shows the energy-band diagram of p-n junction in thermal equilibrium and under forward bias.

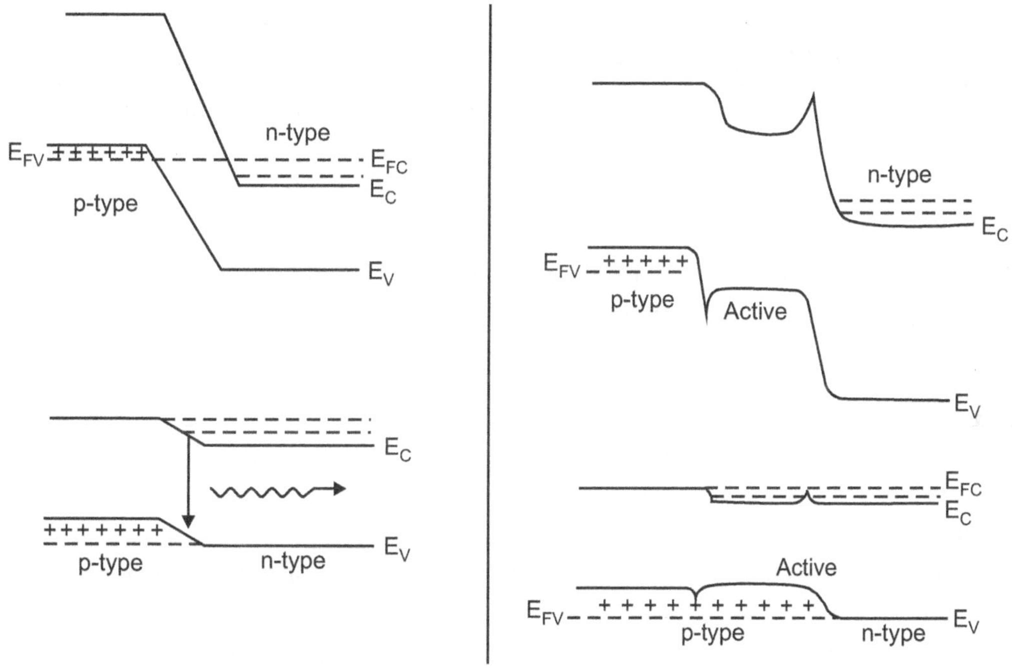

(a) Homostructure Energy Bond Diagram

(b) Double Heterostructure p-n Junction in Top and Bottom

Fig. 3.4

- When a p-n junction is forward biased by applying an external voltage, the built in electric field is reduced. This reduction results in diffusion of electrons and holes across the junction.
- An electric current begins to flow, as a result of carrier diffusion. The current I increases exponentially with applied voltage v.
- Fig. 3.4 (a) is called the **homojunction**, because the some semiconductor material is used on both sides of junction. In a region [see Fig. 3.4 (a)] surrounding the junction also known as depletion width, electrons and holes are present simultaneously when the p-n junction is forward biased.
- These electrons and holes can recombine through spontaneous or stimulated emission and generate light in a semiconductor optical source.
- The problem with homojunction is that, electron-hole recombination occurs over a relatively wide region determined by the diffusion length of electrons and holes.

- Since, the carriers are not confined to the immediate vicinity of the junction, it is difficult to realize high carrier densities. This carrier-confinement problem can be solved by sandwiching a thin layer between p-type and n-type layers, such that the bandgap of the sandwiched layer is smaller than layers surrounding it.
- The carrier confinement occurs as a result of bandgap discontinuity at the junction between two semiconductors which have same structure but different bandgaps. Such junctions are called **heterojunctions** and such devices are called double **heterostructures**.

3.4 NON-RADIATIVE RECOMBINATION

- Non-radiative recombination is a term related to p-n junction. When p-n junction is forward biased, electrons and holes are injected into the active region, where they recombine to produce light. **In any semiconductor, electrons and holes can also recombine non-radiatively**.
- Non-radiative recombination includes, recombination at defects, surface recombination. The radiative and non-radiative recombination times vary from semiconductor to semiconductor.
- The radiative recombination can be written as,

$$\boxed{R_{rr} = R_{spon} + R_{stim}}$$

where, radiative recombination occurs through spontaneous as well as stimulated emission.
- For LEDs, R_{stim} is negligible compared with R_{spon} and R_{rr}.

3.5 SEMICONDUCTOR MATERIALS

- The particular properties of the semiconductor are determined by the materials used and the layering of materials within the semiconductor.
- Silicon (Si) and gallium arsenide (GaAs) are the two, most common semiconductor materials used in electronic and electro-optic devices.
- In some cases, other elements such as aluminium (Al), Indium (In) and phosphorus (P) are added to the base semiconductor materials to modify the semiconductor properties. These elements are called **dopents**.
- Current flowing through a semiconductor optical source causes it to produce light.

3.6 LED

- An LED is a p-n junction diode usually made from a semiconductor material such as aluminium-gallium-arsenide (AlGaAs) or gallium-arsenide-phosphide (GaAsP).

Working Principle:

- LEDs emit light by spontaneous emission, light emitted as a result of the recombination of electrons and holes.
- When current is passes through the PN structure of LED, electrons are injected into an active layer. Light is generated when these injected electrons return from their excited (higher) energy levels to their normal (lower) energy levels. The energy difference is releases as light.

Construction:

- The construction of the LED is shown in Fig. 3.5.
- In the Fig. 3.5, the surface layer is p–type.
- A shallow p-n junction is formed and electrical contacts made to both regions.
- The upper part of the p-material is left uncovered so that radiations from the device are not much affected.
- The inner quantum efficiency of the LED material, sometimes is nearly 100%.
- The external efficiency is much lower.

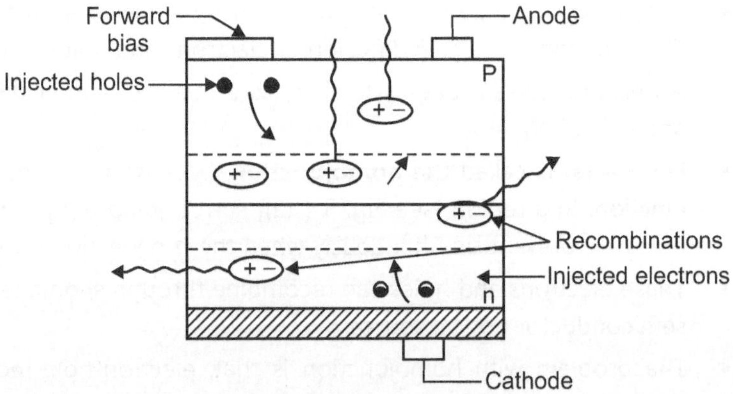

Fig. 3.5: LED Construction

- This is so since most of the emitted radiation strikes the material interface at an angle which is greater than the critical angle and hence, gets trapped.
- The semiconductor/air interface in the LED has the shape of a hemisphere.
- This is made so such that most of the rays strike the surface at less than the critical angle.

Working:

- When forward biased, minority carriers are injected across the P-N junction.
- Once across the junction, these minority carriers recombine with majority carriers and give up energy in the form of light.

Types of LED:

1. Surface emitter LED.
2. Edge emitter LED.
3. Super luminescent LED.
4. Planar LED.
5. Dome LED.

Advantages:

As already stated, LED has a number of advantages. They have –

(a) very high speed (in n sec)
(b) different colours
(c) long life
(d) low voltage operation
(e) low cost
(f) linearity of radiant power output with forward current over a wide range.
(g) low voltage operation to make it compatible with ICs.

Disadvantages:

However, it has some outstanding disadvantages.

(a) It is sensitive to over voltage and over current due to which it is easily damaged.
(b) The radiant output power and wavelength are temperature dependent.
(c) It is difficult to get the theoretical overall efficiency except by using special cooled or pulse conditions.
(d) Large power dissipation.

Applications:

1. They are used in 7-segment, 16-segment and dot-matrix displays, which are used to indicate alphanumeric characters and symbols.
2. They are used for indicating power ON/OFF conditions.
3. They are used in burglar alarm systems.
4. They are used to indicate digital logic state.
5. They can be used as solid state video displays in place of CRTs.
6. They are used in image sensing circuits in picture phones.

Colours:

- By using different LED materials and different dopants, it is possible to get different colours.
- LEDs may be made from gallium phosphide (GaP), gallium arsenide (GaAs) and silicon carbide (SiC). SiC gives yellow light, gallium arsenide phosphide gives red light while gallium arsenide gives infrared radiation. GaP with nitrogen gives green light.

3.6.1 Power-Current Characteristics of LED

- It is easy to estimate the internal power, generated by spontaneous emission.
- At a given current I, the carrier-injection rate is I/q. In the steady state, the rate of electron-hole pairs recombining through radiative and non-radiative process is equal to the carrier injection rate I/q.
- The internal optical power is given by,

$$P_{int} = \eta_{int}\left(\frac{h\omega}{q}\right)I \qquad \ldots (1)$$

where, $h\omega$ = Photon energy

η_{int} = Internal quantum efficiency

- Emitted power is given by,

$$p_e = \eta_{ext} \, p_{int} = \eta_{ext} \, \eta_{int} \left(\frac{h\omega}{q}\right) I \quad \text{... (2)}$$

where, η_{ext} = External quantum efficiency

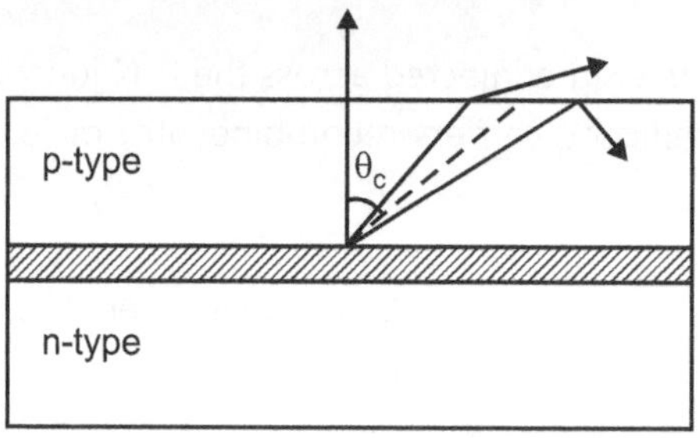

Fig. 3.6: Total Internal Reflection at Output of LED

- As shown in Fig. 3.6, only light emitted within cone of angle θ_c, escapes from the LED surface.

where, $\theta_c = \sin^{-1}\left(\frac{1}{n}\right)$ = Critical angle

n = Refractive index of semiconductor material

- The external quantum efficiency is given by,

$$\eta_{ext} = \frac{1}{4\pi}\int_{\theta}^{\theta_c} T_f(\theta)\,(2n \sin\theta)\, d\theta \quad \text{... (3)}$$

- where, we have assumed that the radiation is emitted uniformly in all directions over a solid angle of 4π.
- The T_f depends upon incidence angle θ. In case of normal incidence, $\theta = 0$, $T_f(0) = 4n/(n + 1)^2$, η_{ext} is approximately by,

$$\eta_{ext} = n^{-1}(n + 1)^{-2} \quad \text{... (4)}$$

- The internal quantum efficiency is generally temperature dependent because of an increase in non-radiative recombination rates at high temperatures.

3.6.2 Modulation Response

- The modulation response of LEDs depends on carrier dynamics and is limited by the carrier lifetime τ_c. It can be determined by using a rate equation for carrier density **N**.
- For LEDs,

$$\frac{dN}{dt} = \frac{I}{qv} - \frac{N}{\tau_c} \quad \text{... (5)}$$

where, the last term includes, both radiative and non-radiative recombination processes through the carrier lifetime τ_c.

3.7 SEMICONDUCTOR LASERS

- **Semiconductor lasers emit through stimulated emission**. As a result of the fundamental difference between spontaneous and stimulated emission.
- They are not only capable of emitting high powers, but also have other advantages related to the coherent nature of emitted light.
- Semiconductor lasers can be modulated directly at high powers, but also have other advantages related to the coherent nature of emitted light.
- Semiconductor lasers can be modulated directly at high frequencies (upto 25 GHz) because of a short recombination time associated with stimulated emission.
- Most fiber-optic communication systems use semiconductor lasers as an optical source, because of their superior performance compared with LEDs.

Construction:

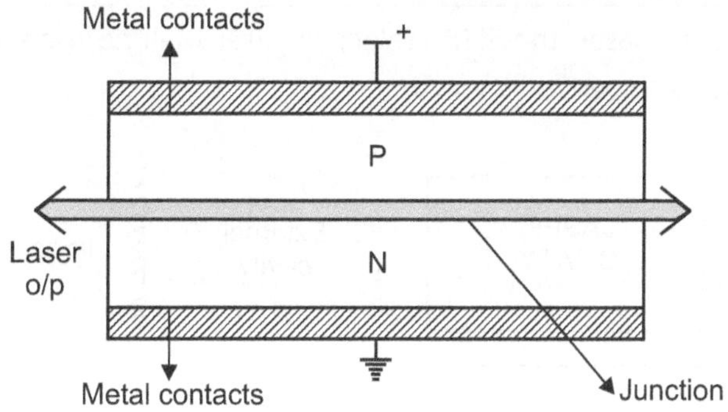

Fig. 3.7: Semiconductor Laser

Working Principle:

- At the junction, light is emitted when electrons or current pass from N to P type material.
- That means, the current is injected into the junction between N and P type materials.

3.8 DISTRIBUTED FEEDBACK LASERS (DFB)

- DFB semiconductor lasers were developed during 1980's.
- The aim of distributed feedback laser is to sharpen up the output of regular Fabry-Perot lasers.

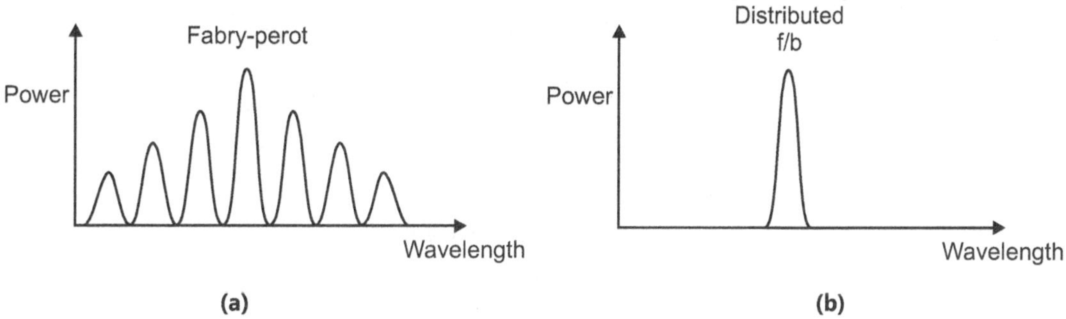

Fig. 3.8

- The feedback in DFB lasers is not localized, but is distributed throughout the cavity length. This is achieved through an internal built-in grating that leads to a periodic variation of the mode index.

Structure:

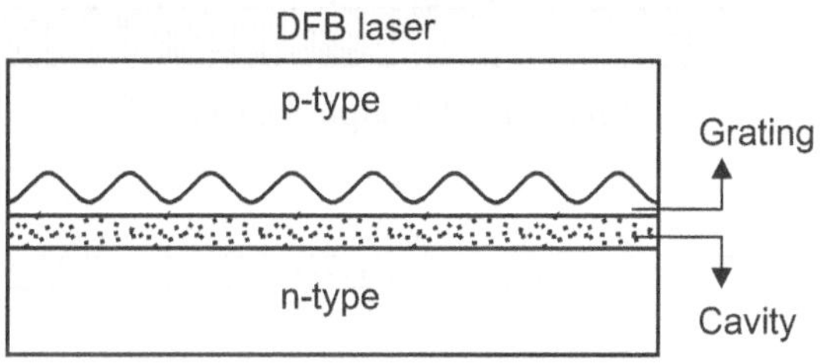

Fig. 3.9: Structure of DFB Lasers

Working Principle:

- As the name suggests, they are feeding back the desired wavelength into the cavity, and this is taking place over the whole length of laser.
- Therefore, the feedback is **"distributed"**.
- The only laser light that builds-up within the active layer is light of this specific wavelength and so that is the only wavelength emitted from laser.

3.9 COUPLED-CAVITY LASERS

- In coupled-cavity semiconductor laser, the SLM (Selective Laser Melting) operation is performed by coupling the light to an external cavity as shown in Fig. 3.10.

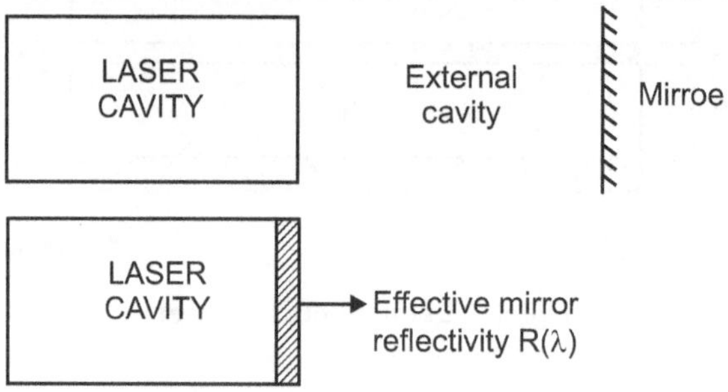

Fig. 3.10: Coupled Cavity Laser

- A portion of the reflected light is feedback into the laser cavity. The feedback from external cavity is not necessarily in phase with optical field inside the laser cavity because the phase shift occurring in the external cavity.
- Several kinds of coupled-cavity schemes have been developed for making SLM laser.

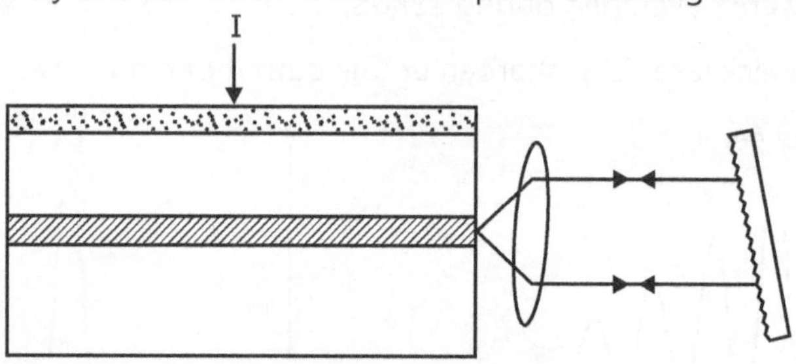

(a): External Cavity Laser

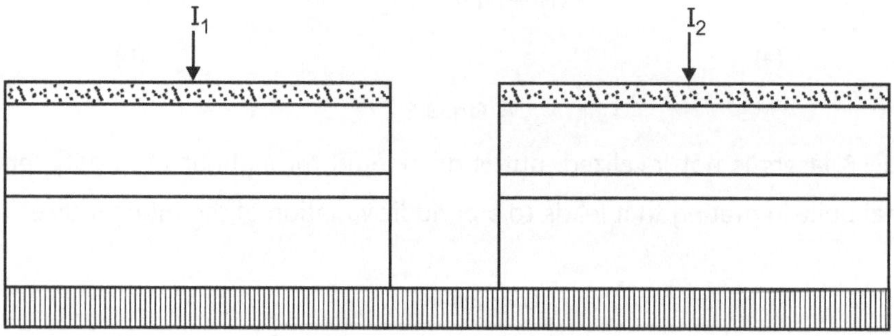

(b) Cleaved-Coupled Laser

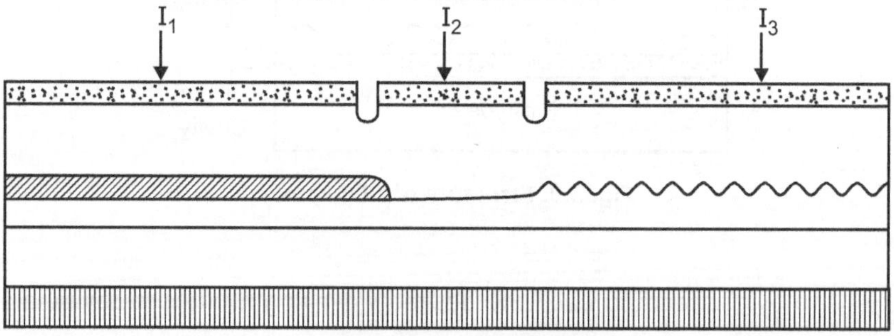

(c) Multisection DBR Laser

Fig. 3.11

- It is necessary to reduce the natural reflectivity of the cleaved facet facing the grating through an antireflection coating to provide a strong coupling. Such lasers are called **external-cavity semiconductor lasers**.

3.10 VERTICAL CAVITY SURFACE EMITTING LASERS (VCSEL)

Working:

- VCSEL operate in a single longitudinal mode virtue of an extremely small cavity length.
- They emit light in a direction normal to the active-layer plane in a manner analogous to that of a surface-emitting LED.
- The emitted light is in the form of circular beam, that can be coupled into a single-node fiber with high efficiency.

Structure:

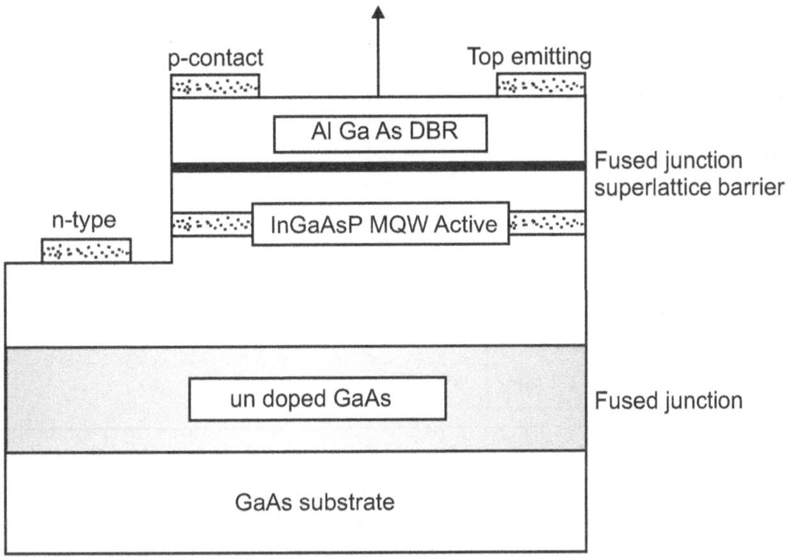

Fig. 3.12: Structure of VCSEL

- As shown in Fig. 3.12, fabrication of VCSEL requires growth of multiple thin layers on a substrate.

Advantages:

1. Lower power consumption.
2. Low beam divergence.
3. On-wafer testing capability.
4. High-fiber coupling efficiency.

3.11 LASER CHARACTERISTICS

- **Laser stands for light amplification by stimulated emission of radiation.**
- Laser is a device which emits a powerful, monochromatic beam of light. The emitted light waves are coherent in nature.
- The characteristics of laser are as follows :

1. **Directionality:**
   - Ordinary light spreads in all directions and its angular spread is 1 m/m. But it is found that laser is highly directional and is angular spread is 1 mm/m.
   - Divergence or angular spread is given by,

$$\phi = \frac{r_2 - r_1}{d_2 - d_1}$$

where, d_1, d_2 = Any two distances.

r_1, r_2 = Radii of beam spots at a distance d_1 and d_2.

2. **Intensity:**
   - Since the ordinary light spreads in all directions, the intensity reaching the target is very less.
3. **Monochromatic:**
   - Laser beam is highly monochromatic. The wavelength is single, whereas in ordinary light, like mercury vapour lamp, many wavelengths of light are emitted.
4. **Coherence:**

- It is an important characteristic of laser beam. In lasers, the wave trains of some frequency are in phase, the radiation given out is in mutual agreement not only in phase, but also in the direction of emission and polarization.
- Thus, it is a coherent beam. Due to this, it results in an extremely high power.

3.12 SMALL SIGNAL MODULATION

- In order to check the improvement in the emitted light signal as a result of matching, the small-signal modulation responses of the laser for both the unmatched and matched cases where plotted as shown in Fig. 3.13.

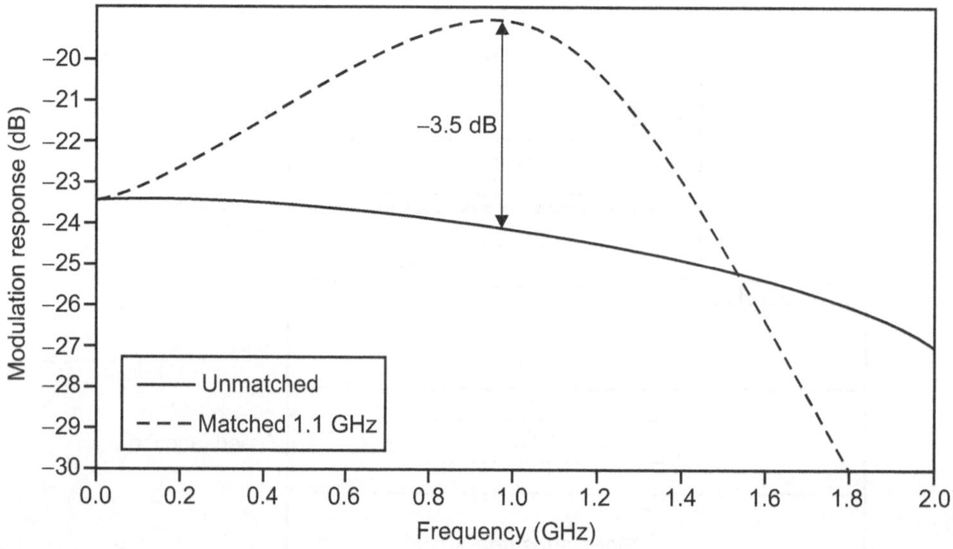

Fig. 3.13: Improvement in the Integrated TLLM Modulation Response, Matched at 1.1 GHz

- **The process of putting information onto a lightwave is called modulation.**
- These results were obtained by applying the FFT to the impulse response of the integrated TLLM model.
- The results shows an improvement in emitted signal of about 3.5 dB. A resonant peak is absent for unmatched case in Fig. 3.13 because the intrinsic laser impedance was simply modelled by resistor (R_d) in parallel with space-charge capacitance (C_{sc}).
- The intrinsic and extrinsic relaxation oscillation frequencies were found to be 2.4 GHz and 1.5 GHz respectively.

3.13 SPECTRAL LINEWIDTH

- In non-semiconductor lasers, for example in solid state lasers, it can be shown that, noise arising from spontaneous emission effects results in a finite spectral width or linewidth ΔV for the lasing output.
- In a semiconductor material, both the optical gain and refractive index depend on the actual carrier density in the medium.
- This relationship leads to an index-gain coupling mechanism; that is, it gives rise in an interaction between phase noise and light intensity.

$$\Delta v = \frac{R_{sp}}{4nI}(1 + \alpha^2) \quad \ldots (1)$$

where,

I = Average number of photons in lasing cavity

R_{sp} = Spontaneous emission rate

α = Linewidth enhancement factor

- The linewidth expression in equation (1) can be rewritten in terms of optical power output as,

$$\Delta v = \frac{v_g^2\, hvg_{th}\, n_{sp}\, \alpha_t}{8\pi P_{out}}(1 + \alpha^2) \quad \ldots (2)$$

where,

v_g = Group velocity of light.

hv = Photon energy.

g_{th} = Threshold gain.

α_t = Cavity loss.

n_{sp} = Spontaneous emission factor.

- The spectral linewidth of laser can also increase significantly when direct modulation is used to vary the light output level.

3.14 SOURCE FIBER COUPLING

- In source fiber coupling two approaches have been used as follows:
  1. Direct or Butt coupling
  2. Lens-coupling.

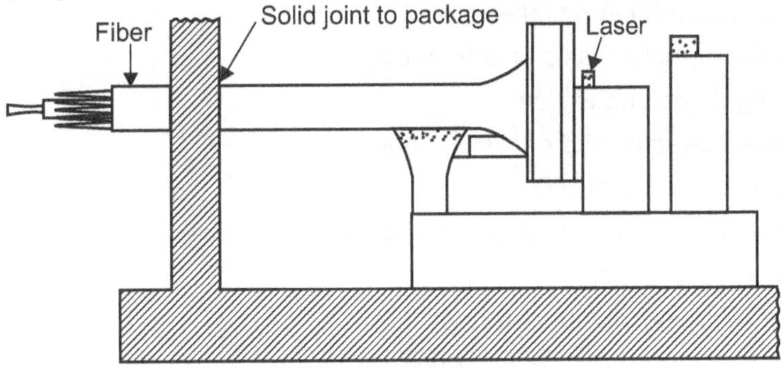

Fig. 3.14 (a): Direct Coupling

- In Butt-coupling, the fiber is brought close to the source and held in place by epoxy.
- The coupling of a semiconductor laser to a single-mode optical fiber is more efficient than that of an LED. The butt coupling provides only about 10% efficiency.
- Typically index-guided InGaAsP lasers have a mode size of about 1 μm, whereas the mode size of a single-mode fiber is in the range of 6-9 μm.

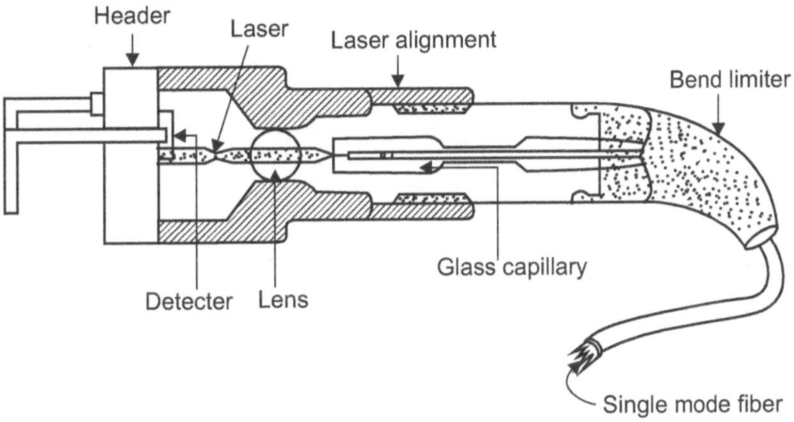

Fig. 3.14 (b): Lens-Coupling

- As shown in Fig. 3.14 (b), a lens is used to maximize the coupling efficiency. The coupling efficiency can exceed 70%. The alignment of the fiber core is less critical for the confocal design because the spot size is magnified to match the fiber's mode size.
- Each approach has its own merits and choice, generally depends on the design objectives.

Important Points

- Emission process occurs in two ways:
  1. By spontaneous emission, 2. By stimulated emission.
- An LED is an p-n junction diode, made from semiconductor material such as AlGaAs or GaAsP.
- **LEDs emit light by spontaneous emission, light emitted as a result of the recombination of electrons and holes.**
- When some material is used on both the sides of junction then it is called as **homojunction**.
- When the semiconductor is having some structure but different bandgaps, then such junctions are called as **heterojunctions**.
- Semiconductor lasers emit light through **stimulated emission**.
- Source fiber coupling contains two approaches as follows :
  1. Direct or Butt coupling, 2. Lens-coupling.

Practice Questions

1. Define emission in laser.
2. Draw p-n junction and explain.
3. How non-radiative recombination takes place?
4. Explain semiconductor materials used in fiber optic.
5. Draw light emitting diode and explain.
6. Give power current characteristics of LED in detail.
7. Explain modulation response of LED.
8. Give different types of semiconductor lasers and explain any one.
9. Explain DFB laser with diagram.
10. Describe coupled cavity semiconductor laser.
11. How vertical cavity semiconductor laser works?
12. Give laser character in detail.
13. Explain small signal modulation.
14. Define spectral linewidth.
15. Explain any one type of source fiber coupling.

***

Chapter 4

OPTICAL RECEIVERS

4.1 INTRODUCTION

- The role of an optical receiver is to convert the optical signal back into electrical form and recover the data transmitted through the lightwave system.
- The main component is photodetector that converts light into electricity through the photoelectric effect.
- In this chapter, p-n photo diodes, PIN photo diodes, Avalanche photo diode, MSM photo detector are described.

4.2 P-N PHOTODIODE

Definition:

- **A special type of P-N junction device that generates current when exposed to light is known as photodiodes.**
- It is also known as **photodetector** or **photosensor**.
- It operates in reverse biased mode and **converts light energy into electrical energy**.
- It works on the principle of **photoelectric effect**.

Fig. 4.1: Symbol of Photodiode

Construction of p-n Photodiode:

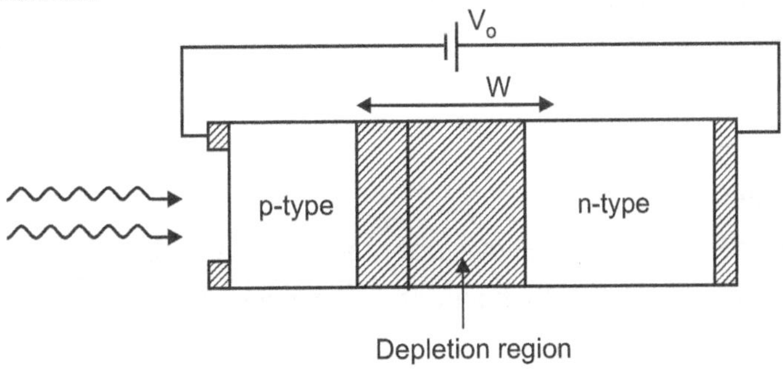

Fig. 4.2: A p-n Photodiode under Reverse Bias

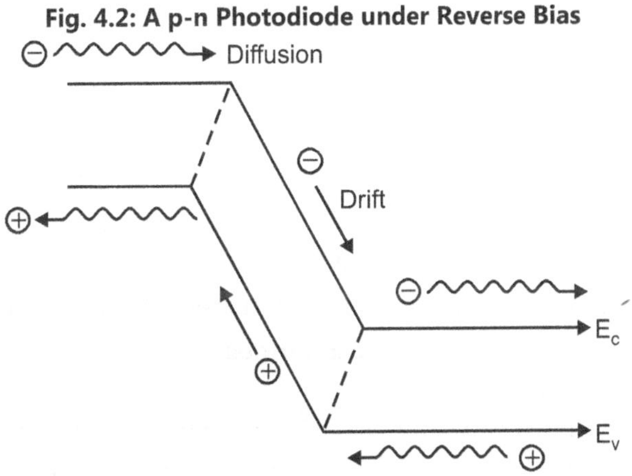

Fig. 4.3: Energy Bond Diagram

Working:

- A reverse biased p-n junction consists of a region known as the **depletion region**, that is essentially devoid of free charge carriers and where a large built in electric field opposes flow of electrons from the n-side to p-side and holes from P to n.
- When such a p-n junction is illuminated with light on one side, the electron-hole pairs are created through absorption. Because of the large built-in electric field, electrons and holes generated inside the depletion region accelerate in opposite directions and drift to the n and p-side respectively.
- The resulting flow of current is proportional to the incident optical power. Thus, a reverse-biased p-n junction acts as a photodetector and is known as **p-n photodiode**.

Characteristics of Photodiode:

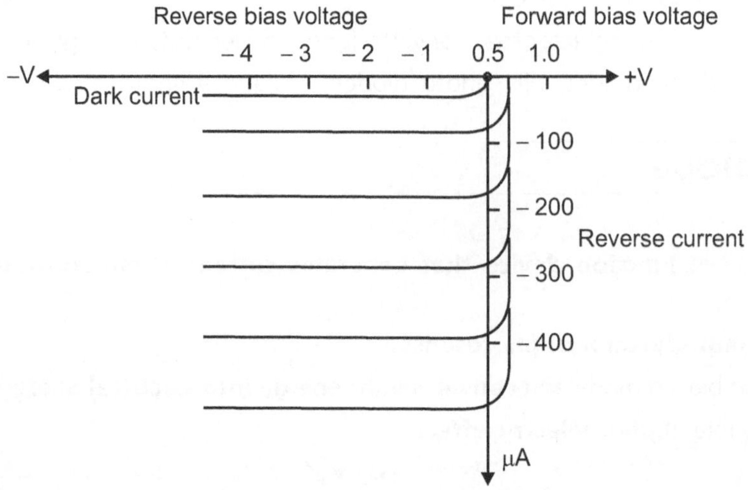

Fig. 4.4: Characteristics of Photodiode

Advantages of Photodiode:

1. Shows quick response, when exposed to light.
2. Low cost device.
3. Offers high operational speed.
4. Provides linear response.

Disadvantages of Photodiode:

1. Temperature-dependent device and shows poor temperature stability.
2. When low illumination is provided, then amplification is necessary.

Applications of Photodiode:

1. Counters and switching circuits.
2. Optical communication systems.
3. Logic circuits and encoders.
4. Alarm systems.

4.3 PHOTO DIODES

- Photo diodes convert light directly to electric current.
- **PIN diode can convert one photon to one electron of current. Hence, amplifier is needed with receiver.**

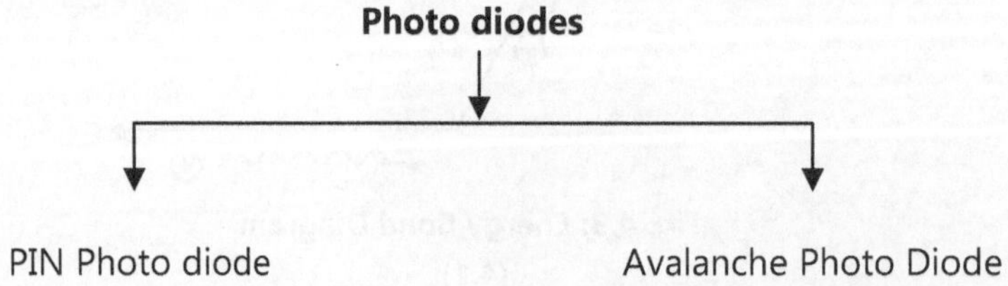

4.3.1 PIN Diodes

Name PIN because intrinsic layer introduced between P and N region.

Principle:

Principle of PIN diodes is reverse that of LED. That is, light is absorbed at p-n junction rather than emitted.

The wide intrinsic i layer has only small degree of doping and has wide depletion layer. There are number of modifications:

1. It increases the chances of entering photon being absorbed, because amount of absorbant material is more.
2. Since junctions are wider, junction capacitance is less.
3. Two ways of current flow are diffusion and drift.

Thus i layer increases responsivity and decreases response time.

Construction:

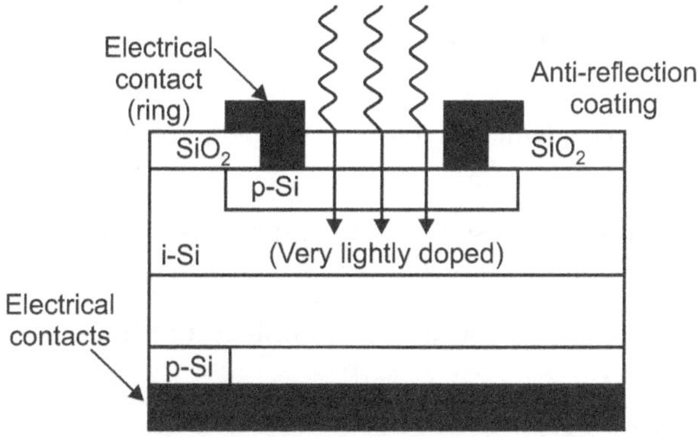

(a) Typical Silicon PIN Diode Schematic

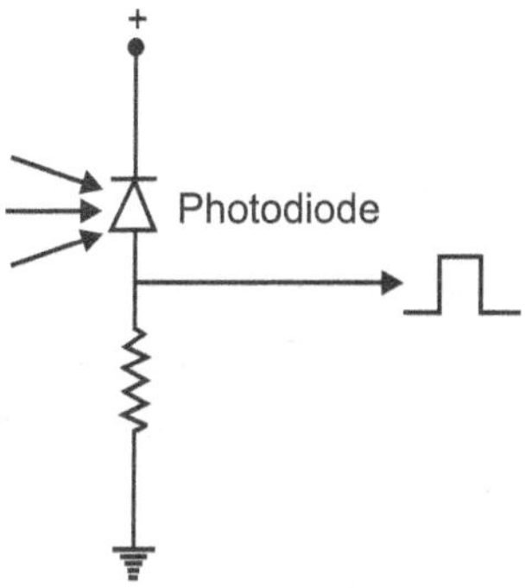

(b) How diode converts light into voltage pulses

Fig. 4.5

Working:

- The sensitivity and response time can be increased by adding an undoped intrinsic (I) layer between the P and N semiconductor to form a PIN diode.
- The I layer exposed to light.
- This is a silicon PN junction diode that is sensitive to light.
- This diode is normally reverse-biased.
- The only current that flows through it is an extremely small leakage current.
- Whenever light strikes the diode, this leakage current will increase significantly.
- It flows through a resistor and develops a voltage drop across it.
- The result is an output voltage pulse.
- Voltage pulse is small, so that it must be amplified.
- For operating PIN diode, energy of absorbed photon must be sufficient to promote an electron across bandgap.
- At energy higher than bandgap energy, material absorbs photons.
- Germanium is material used as it has low bandgap energy regardless of wavelength.

Characteristics of PIN diodes:

1. **Quantum efficiency:**

It is defined as the ratio of number of electrons collected at the junction over the number of incident photons. PIN diodes has 1 electron release per 1 photon.

2. **Responsivity:**

It is a measure that does take photon energies into account. It is output photo current of device divided by the input optical power.

Typical responsivity for λ = 900 nm is 0.44.

4.3.2 Avalanche Photo Diodes (APDs)

- APDs amplify the signal during the detection process. They use a similar principle to that of photomultiplier tubes used in nuclear detection.

Principle:

A single photon acting on the device releases a single electron. This electron is accelerated through an electric field until it strikes a target material.

This collision releases multiple electrons. These electrons get accelerated and strike another target.

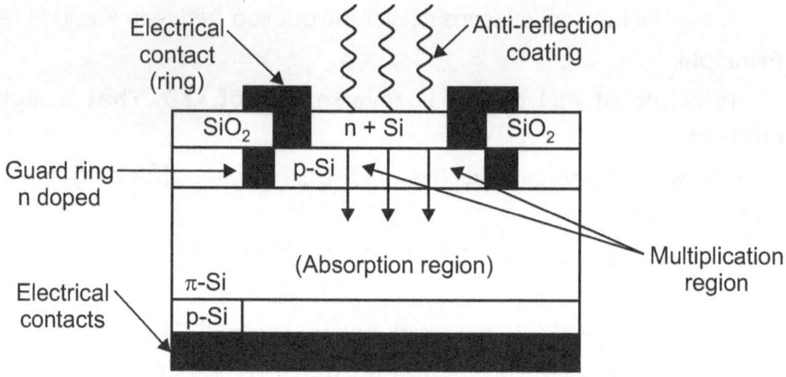

Fig. 4.6: Avalanche Photo Diode (APD)

This releases more electrons. This process is repeated. Thus, one electron has created large current.

Working:

- Structural difference between APD and p-i-n diode is that i zone (which is lightly n-doped in p-i-n structure) is lightly p-doped and renamed π layer.
- Arriving photons generally pass straight through n^+-p junction (because it is thin) and are absorbed in π layer.
- This absorption produces free electron.
- Electric potential across π layer is sufficient to attract the electron towards one contact and holes at other contact.
- Electrons are attracted towards n^+ layer at top of the device.
- Around the junction between n^+ and p-layers the electric field is so intense that carriers are strongly accelerated.
- When these electrons collide with other atom, they produce electron-hole pair.
- This process is called as impact ionization.
- When the light strikes the junction, breakdown occurs and large current flows.
- This high reverse current requires less amplification.
- Like PIN diode APD also reverse biased.
- When a sufficient amount of reverse will flow due to avalanche effect.

Avalanche Photodiode (APD) Char:

1. **Sensitivity of APDs:**

   Very high sensitivity makes APD very useful.

2. **Gain bandwidth product:**

   Performance of optical detector is decided by its gain bandwidth product.

   APD has gain bandwidth product of 150 GHz.

3. **Operating speed:**

   Device capacitance limit speed of p-i-n diodes.

   APD takes too long to make impact ionization effect.

   Hence, maximum speed of APD is restricted.

Advantages:

- It is fastest and most sensitive photodiode.
- Capable to handle very high gigabit/sec. data rates possible in some systems.

Disadvantage:

- It is expensive and circuitry is complex.

4.4 METAL SEMICONDUCTOR METAL (MSM) PHOTODIODE

- An MSM structure consists of two shottky electrodes, often interlinked in the form of comb structure, leaving a fee semiconductor surface between the two contacts which forms the active region in which light will be absorbed.
- A bias voltage can be applied between the two electrodes, in order to break the initial electrical symmetry of contacts. One of the schottky junction is reversed-biased, producing a SCR of increased width, and the other junction is forward biased.
- The absorption of light near the reverse-biased junction creates electron-hole pairs, which are separated under the effects of the electric field present in the SCR, thus creating the photocurrent.
- The other electrode, consisting of a forward-biased schottky junction, simply acts as a collection electrode.
- The band diagram of the device under increased bias voltage (v_B) is presented in Fig. 4.7 which L is the distance between two adjacent contact figures, ϕ_0 is the height of schottky barrier and I_{ph} is the photocurrent.

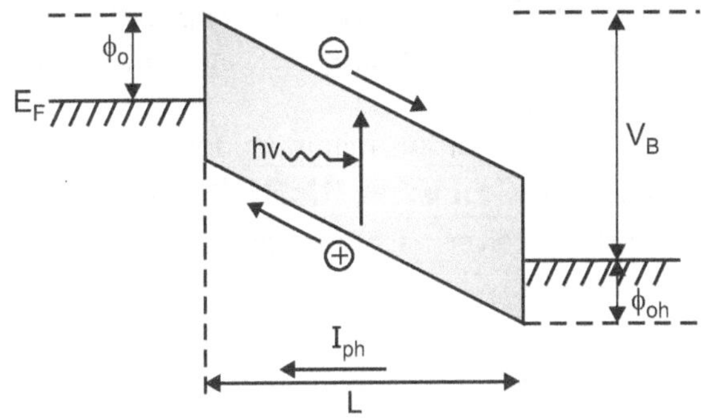

Fig. 4.7: Energy Band Diagram of MSM Photodiode

4.5 RECEIVER DESIGN

Fig. 4.8 shows the basic components of an optical receiver.

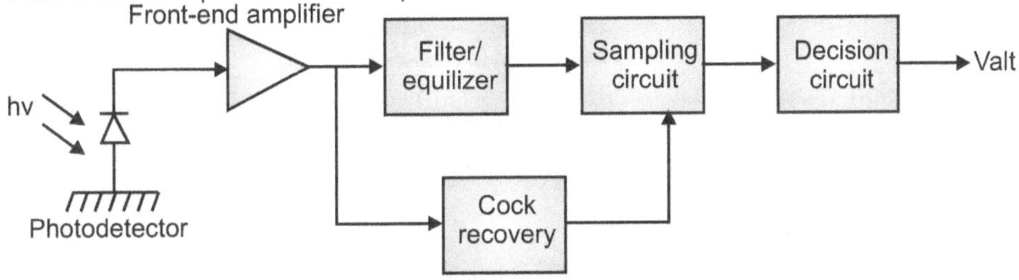

Fig. 4.8: Basic Sections of Optical Receiver

Explanation:

- **Sampling Decision Circuit:**
  It samples, the signal level at the mid-point of each time slot and compares it with a certain reference voltage known as **threshold voltage**.
- **Clock Recovery:**
  - If the received signal level is greater than the threshold level, 1 is said to have been received.
  - If the voltage is below the threshold level, 0 is assumed to have been received.
  - To accomplish this bit interpretation, the receiver must know, where the bit boundaries are.
  - This is done with the periodic waveform called **clock** and the function is called **clock recovery** or **timing recovery**.
- **Photodetector:**
  Produces an electric current that is proportional to the received power level.
- **Front-end Amplifier:**
  Boosts the weak electric current.
- **Equilizer:**
  Equilizes or cancels pulse-spreading effects.

4.6 RECEIVER NOISE

- The term **noise** is used to describe unwanted components of an electrical signal that tend to disturb the transmission and processing signal in physical system.
- Receiver noise includes thermal noise, dark current noise, quantum noise.

Thermal Noise:

- It is the noise resulting from the random motion of electrons in a conducting medium.
- Thermal noise carries from both photodetector and load resistor.

Shot Noise (Quantum Noise):

- Shot noise is the noise caused by current fluctuations because of the discrete nature of charge carriers.

Dark Current Noise:

- It results from dark current that continues to flow in the photodiode when there is no incident light.

Thermal Noise

- It arises from the random motion of electrons in a conductor.

4.7 RECEIVER SENSITIVITY

- Optical communication systems use a BER (Bit Error Rate) value to specify the performance requirements for a particular transmission link application.
- To achieve a desired BER at a given data rate, a specific minimum average optical level must arrive at the photodetector.
- The value of this minimum power level is called **receiver sensitivity**.

4.8 RECEIVER PERFORMANCE

- The receiver performance is characterized by measuring the BER as a function of the average optical power received. The average optical power corresponding to a BER of 10^{-9} is a measure of receiver sensitivity.
- The receiver performance is generally better for shorter wavelengths in the region near 0.85 μm, where silicon APD's can be used.
- The performance of an optical receiver in actual lightwave systems may change with time. Since it is not possible to measure BER directly for a system in operation.
- The receiver performance can also be improved by amplifying the optical signal before it is incident on the photodetector.

Important Points

- An LED is a p-n junction diode usually made from a semiconductor material such as aluminium-gallium-arsenide (AlGaAs) or gallium-arsenide-phosphide (GaAsP).
- LEDs emit light by spontaneous emission, light emitted as a result of the recombination of electrons and holes.
- **Principle:** Laser action is the result of three process absorption of energy packets (photons), spontaneous emission, and stimulated emission. (These processes are represented by the simple two-energy-level diagrams).
- Predominant types of receiver used in communication system rely on principle of ionization of semiconductor.
- PIN diode can convert one photon to one electron of current. Hence, amplifier is needed with receiver.

Photo Diodes

PIN Photo diode | Avalanche Photo Diode

- **Principle:** Principle of PIN diodes is reverse that of LED. That is, light is absorbed at p-n junction rather than emitted.
- APDs amplify the signal during the detection process. They use a similar principle to that of photomultiplier tubes used in nuclear detection.

Practice Questions

1. Explain p-n photodiode with diagram.
2. Give characteristics of photodiode.
3. Give advantages, disadvantages and applications of p-n photodiode.
4. Draw PIN photodiode and explain in detail.
5. Draw avalanche photodiode and explain.
6. Describe MSM photo detector with diagram.
7. Draw block diagram of receiver and explain each bock.
8. Define receiver sensitivity.
9. Give receiver performance in detail.
10. Draw energy bond diagram of MSM photodetector.

***

Chapter 5

LIGHT WAVE SYSTEMS

5.1 INTRODUCTION

- In this chapter, we will see the practical lightwave system which includes architecture, in that first involves point-to-point link.
- It includes examining the components that are available for a particular application.
- Distribution networks involves hub and bus topology for distribution purpose.
- The common topologies in LAN are ring and star topology which consists of shared transmission medium.
- Power budget analysis will determine whether the fiber optic link meets the attenuation requirements or if amplifiers are needed within the link to boost the power level.
- Sources of power penalty includes modal noise, dispersive pulse broadening, frequency chirping and reflection feedback noise.

5.2 LIGHT WAVE SYSTEM ARCHITECTURE

- Fiber optic communication system can be classified into three broad categories as follows :
  (i) Point-to-point links (ii) Distribution networks
  (iii) Local Area Networks (LANs)

Explanation:

5.2.1 Point-to-Point Links

- This is the simplest form of lightwave system. Its function is to transport information that is available in the form of a digital bit stream, from one place to another place, as accurately as possible.

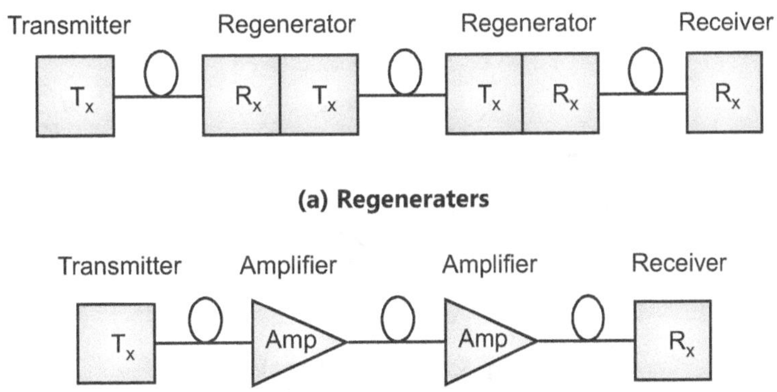

(b) Optical Amplifiers

Fig. 5.1

- Fig. 5.1 shows two concepts used for loss compensation.
- As the name suggests, regenerators, regenerate the optical signal. **Regenerater** is nothing but the receiver transmitter pair that detects the incoming optical signal, recovers the electrical bit stream and then converts it back into optical form by modulating an optical source.
- **Optical amplifiers**, amplifies the optical beat stream directly, without requiring conversion of the signal to the electric domain.

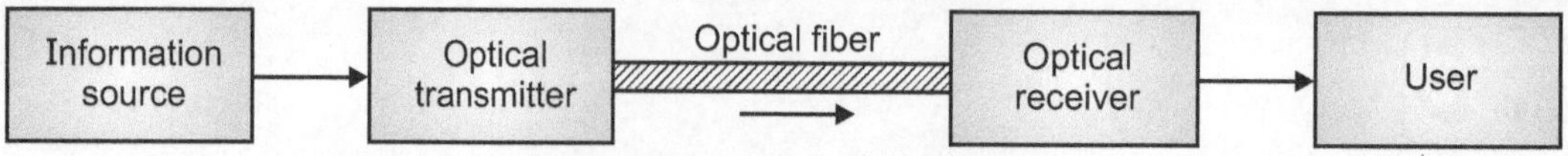

Fig. 5.1 (c): Point-to-Point Link

- The simplest point-to-point link has a transmitter on one end and a receiver on the other end as shown in Fig. 5.1 (c).
- The following key system requirements are needed in analyzing a link:
  1. The desired transmission distance.
  2. The data rate or channel bandwidth.
  3. The bit-error rate (BER).
- To fulfill these requirements, the designer has a choice of the following components and their characteristics.

(a) Multimode or Single-mode Optical Fiber:

1. Core size.
2. Core refractive index profile.
3. Bandwidth or dispersion.
4. Attenuation.
5. Numerical aperture.

(b) LED or Laser Diode Optical Source:

1. Emission wavelength.
2. Spectral line width.
3. Output power.
4. Effective radiating area.
5. Emission pattern.
6. Number of emitting modes.

(c) Pin or Avalanche Photodiode:

1. Responsitivity.
2. Operating wavelength.
3. Speed.
4. Sensitivity.

5.2.2 Distribution Networks

- Many applications of optical communication systems require the information that is not only transmitted but is also distributed to a group of subscribers.
- Fig. 5.2 (a) shows two topologies for distribution networks.
  (i) Hub topology, (ii) Bus topology.

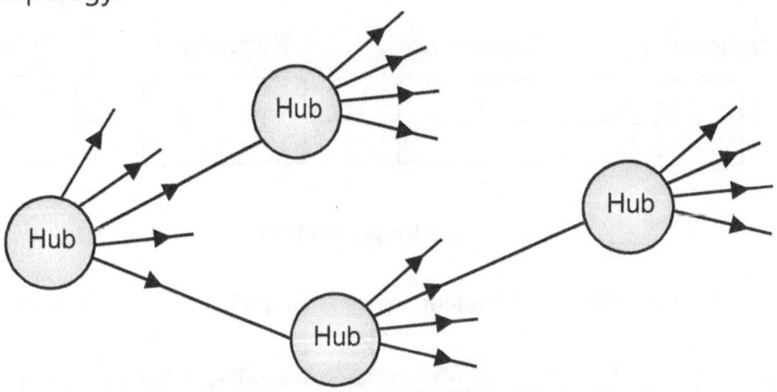

Fig. 5.2 (a): Hub Topology

(i) Hub Topology:

- In this case, the channel distribution takes place at central locations called as **hub**. Such networks are called as **Metropolitan Area Networks (MANs)** as hubs are typically located in major cities.

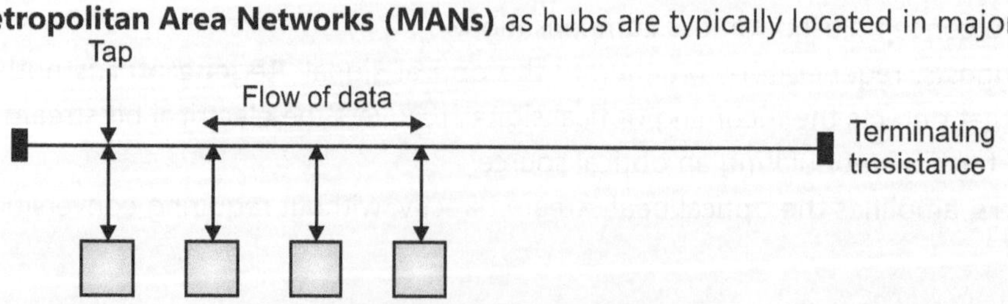

Fig. 5.2 (b): Bus Topology

(ii) Bus Topology:

- In this, all stations attach appropriate hardware interfacing known as a tap, directly to a linear transmission medium or a bus.
- There is full-duplex operation between the station and the tap that allows data to be transmitted onto the bus and received from the bus.
- A transmission from any station propagates the length of the medium in both directions and can be received by all other stations.
- There is a terminator at each end of the bus, which absorbs any signal, removing from the bus.

5.2.3 Local Area Network (LAN)

- A LAN consists of a shared transmission medium and a set of hardware and software for interfacing devices to the medium and regulating the orderly access to the medium.
- The common topologies in LAN are: (i) Ring topology, (ii) Star topology.
- In communication network, the term topology refers to the way in which the ends points or stations, attached to the network are interconnected.

(i) Ring Topology:

- In **ring topology**, consecutive nodes are connected by point-to-point links to form a closed ring. Each node can transmit and receive the data by using a transmitter-receiver pair, which also acts as a repeater.
- A token is passed around the ring, each node monitors the bit stream to listen for its own address and to receive the data.

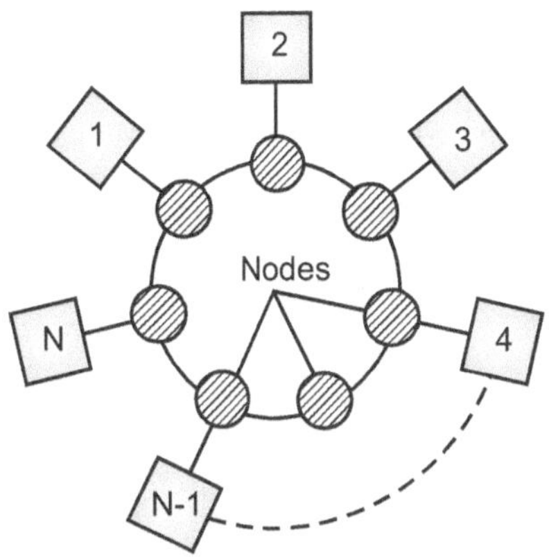

Fig. 5.3 (a): Ring Topology

(ii) Star Topology:

- In **star topology**, each station or node (user) is directly connected to a common central node. Each station attaches to central node through two point-to-point links, one for transmission and one for reception.

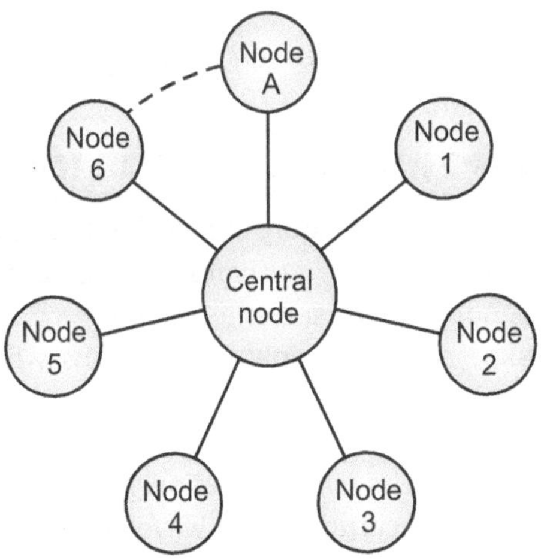

Fig. 5.3 (b): Star Topology

5.3 LOSS-LIMITED LIGHT WAVE SYSTEM

- The design of fiber-optic communication systems requires a clear understanding of the limitations imposed by the loss, dispersion and non-linearity of fiber.
- In this, we will see, how the bit rate and transmission distance of a single-channel system are limited by fiber loss and dispersion.

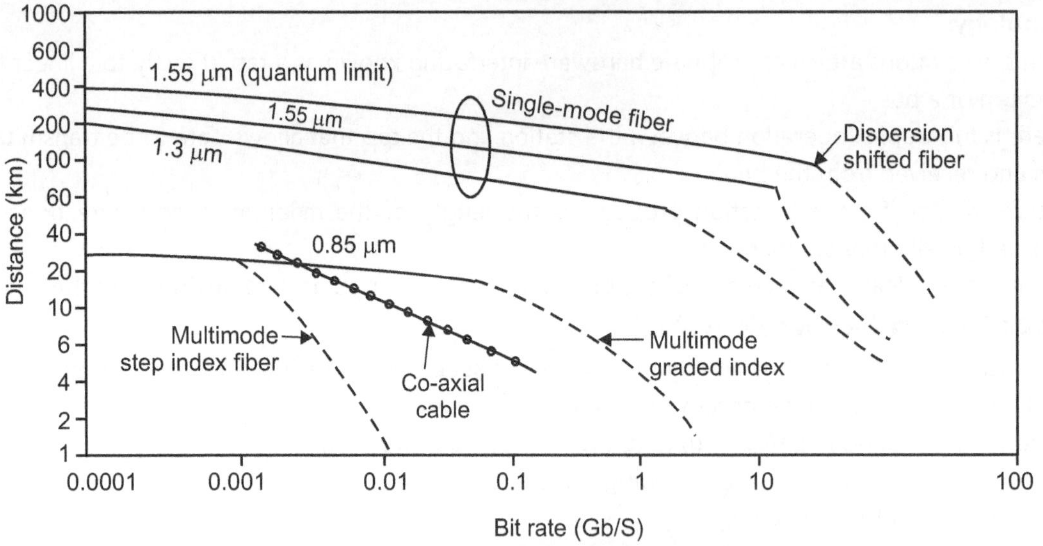

Fig. 5.4: Loss (Solid Line) and Dispersion (Dashed Line) Limits on Transmission Distance L as a Function of Bit Rate B for the Three Wavelength Windows

- Consider an optical transmitter that is capable of launching an average power $\bar{P}_{tr}$. If signal is detected by a receiver that requires a minimum average power $\bar{P}_{rec}$ at the bit rate B, the maximum transmission distance is limited by,

$$L = \frac{10}{\alpha_f} \log_{10}\left(\frac{\bar{P}_{tr}}{\bar{P}_{rec}}\right) \quad \dots (1)$$

where,

α_f = Net loss in dB/km

$\bar{P}_{rec} = \bar{N}_p h\nu B$

$h\nu$ = Photon energy

$\bar{N}_P$ = Average number of photons

- The solid line in figure shows the dependence of L and B for three common operating wavelengths of 0.85, 1.3 and 1.55 μm.

5.4 DISPERSION-LIMITED LIGHTWAVE SYSTEMS

- The dashed line in figure shows the dispersion-limited transmission distance as a function of the bit rate.
- Since the physical mechanisms leading to dispersion limitation can be different for different operating wavelengths, let us see one-by-one.
- In first case, 0.85 μm lightwave systems, which often use multimode fibers to minimize the system cost.
- The second generation of lightwave systems used primarily single-mode fibers near the minimum-dispersion wavelength occurring at about 1.31 μm.
- The third and fourth generation lightwave systems operate near 1.55 μm to take advantage of the smallest fiber losses occurring in this wavelength region.
- A solution to the dispersion problem is offered by dispersion-shifted fibers for which dispersion and loss both are minimum near 1.55 μm.

5.5 POWER BUDGET

- The purpose of power budget is to ensure that, enough power will reach the receiver to maintain reliable performance during the entire system lifetime.
- The minimum average power required by the receiver is the receiver sensitivity $\bar{P}_{rec}$. The average launch $\bar{P}_{tr}$ is generally known as transmitter.
- The power budget is expressed in **dBm**.

$$\bar{P}_{tr} = \bar{P}_{rec} + C_L + M_S \quad \text{... (1)}$$

where,

C_L = Total channel loss

M_S = System margin

- Following table shows the power budget for two transmitters by assuming that the splice loss is included within the cable loss.

Table 5.1: Power Budget of a 0.85 μm Lightwave System

Quantity	Symbol	Laser	LED
Transmitter power	$\bar{P}_{tr}$	0 dBm	−13 dBm
Receiver sensitivity	$\bar{P}_{rec}$	−42 dBm	−42 dBm
System margin	M_S	6 dB	6 dB
Available channel loss	C_L	36 dB	23 dB
Connector loss	α_{con}	2 dB	2 dB
Fiber cable loss	α_f	3.5 dB/km	3.5 dB/km
Maximum fiber length	L	9.7 km	6 km

5.6 LONG HOUL SYSTEMS

- Long houl optics refers to the transmission of visible light signals over optical fiber cable for great distances, especially without or with minimal use of repeaters.
- Normally, repeaters are necessary at intervals in a length of fiber optic cable to keep the signal quality from deteriorating to the point of non-usability.
- In long-houl optical systems, the goal is to minimize the number of repeaters per unit distance.

5.7 SOURCES OF POWER PENALTY

- The sensitivity of the optical receiver in a realistic light wave system is affected by several physical phenomena which is combination with fiber dispersion, degrade the SNR at the decision circuit.
- In this section, we will see how the system performance is affected by fiber dispersion by considering the extent of power penalty.

5.7.1 Model Noise

- Model noise is associated with multimode fibers. Interference among various propagation modes in a multimode fiber creates a **speckle pattern** at the photodetector.
- The non-uniform intensity distribution associated with the speckle pattern is harmless in itself, as the receiver performance is governed by the total power integrated over the detector area.
- However, if the speckle pattern fluctuates with time, it will lead to fluctuations in the received power that would degrade the SNR. **Such fluctuations are referred to as model noise.**

5.7.2 Dispersive Pulse Broadening

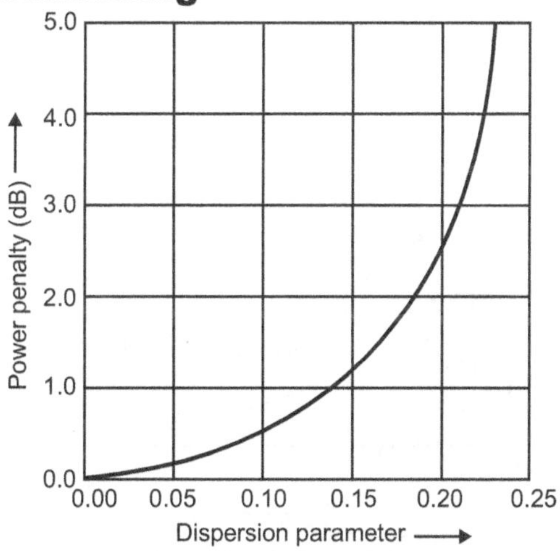

Fig. 5.5: Dispersion Induced Power Penalty

- Dispersion-induced pulse broadening affect the receiver performance in two ways.
- First, a part of the pulse energy spreads beyond the allocated bit slot and leads to intersymbol interference (ISI).
- Second, the pulse energy within the bit slot is reduced when the optical pulse broadens. Such a decrease in pulse energy reduces the SNR at the decision circuit.

5.7.3 Mode Partition Noise

- Multimode semiconductor lasers, exhibit Mode Partition Noise (MPN), a phenomenon occurring because of an anticorrelaton among pairs of longitudinal modes.

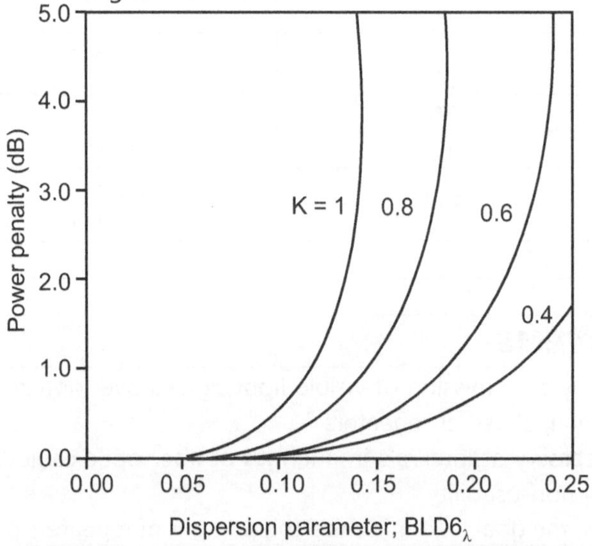

Fig. 5.6: MPN-Induced Power Penalty vs. BLD σ_λ

- Mode-partition noise is associated with intensity fluctuations in the longitudinal modes of multimode laser diode. Intensity fluctuations can occur among the various modes in a multimode laser even when the total optical output is constant.
- Mode-partition noise becomes more pronounced for higher bit rates. The errors due to mode-partition noise can be reduced and sometimes eliminated by setting the bias point of the laser above threshold.

5.7.4 Frequency Chirping

- A laser that oscillates in a single longitudinal mode under CW operation may experience dynamic line broadening, when the injection current is directly modulated. This line broadening is a **"frequency chirp"** associated with modulation-induced changes in the carrier density.
- Laser chirping can lead to significant dispersion effects for intensity-modulated pulses when the laser emission wavelength is displaced from the zero-dispersion wavelength of the fiber.

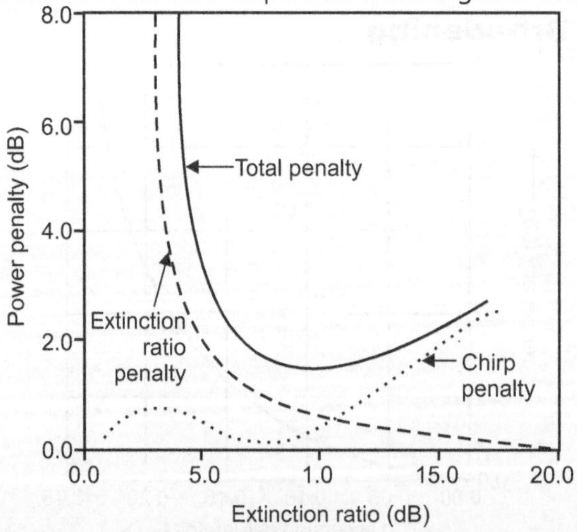

Fig. 5.7: Power Penalty as a Function of Extinction Ratio

5.7.5 Reflection Feedback Noise

- In most fiber optic communication systems, some light is invariably reflected back because of refractive index discontinuities occurring at splices, connectors and fiber ends.
- Even a relatively small amount of optical feedback affects the operation of semiconductor lasers and can lead to excess noise in transmitter output.
- When light travels through a fiber link, some optical power gets reflected as refractive-index discontinuities. The reflected signals can degrade both transmitter and receiver performance.
- In high speed systems, this reflected power causes optical feedback, which can induce laser instabilities. These instabilities show up as either intensity noise (output power fluctuations), jitter (pulse distortion0, or phase noise in the laser, and they can change its wavelength, linewidth and threshold current.
- Since they reduce the SNR, these effects cause two types of power penalties in receiver sensitivities. First, as shown in Fig. 5.8 (a), multiple reflection points set-up an interferometric cavity that feeds power back into the laser cavity, thereby converting phase noise into intensity noise.

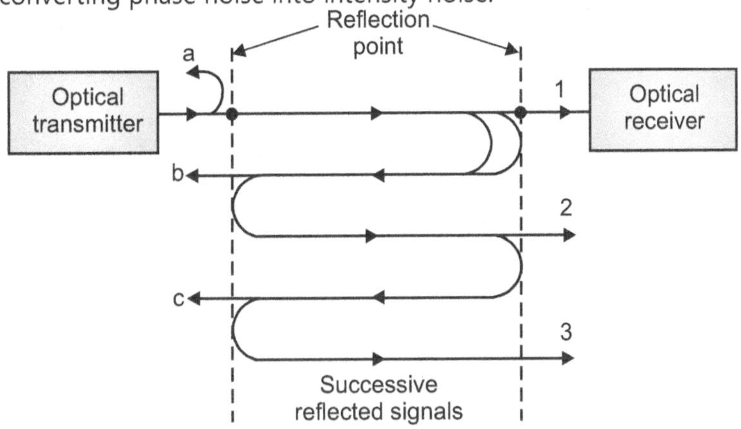

Fig. 5.8 (a): Refractive Index Discontinuities can Setup Multiple Reflections in a Fiber Link

- A second effect created by multiple optical paths in the appearance of spurious signals arriving at the receiver with variable delays, thereby causing intersymbol interference (ISI), as shown in Fig. 5.8 (b).

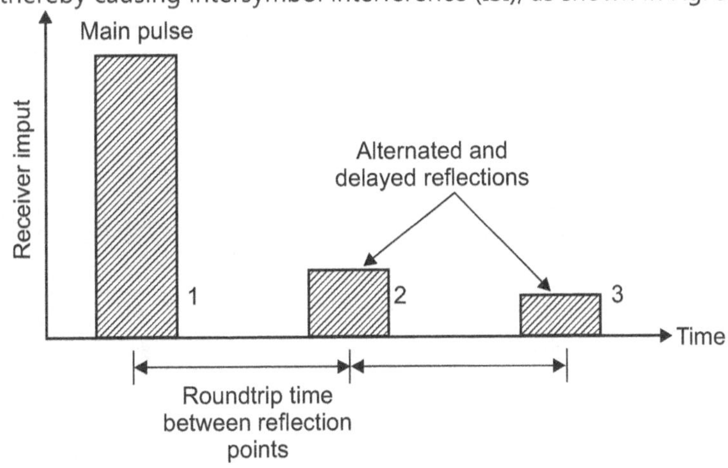

Fig. 5.8 (b); Each Round Trip of a Light Pulse Reflection Creates Another Attenuated and Delayed Pulse, Which Can Cause ISI

Important Points

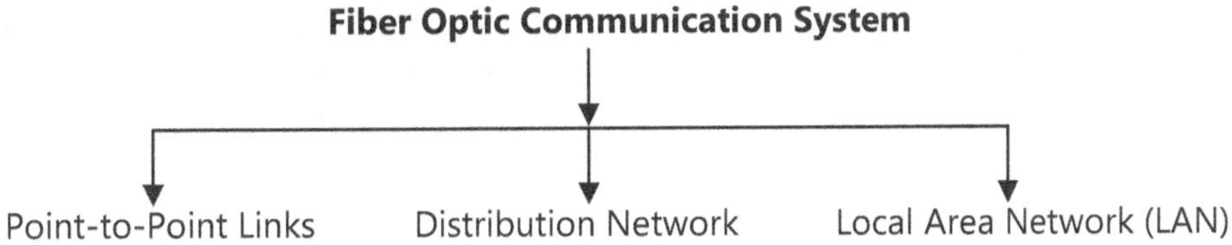

- The simplest form of fiber optic communication is point-to-point link, which has a transmitter on one side and receiver on other side.
- Many applications of optical communication requires the information that is not only transmitted but also distributed to a group of subscribers. In such applications, **distribution networks** are required.

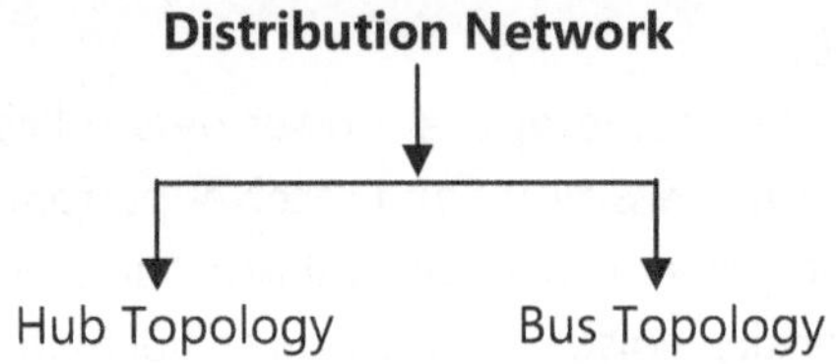

- **Channel distribution takes place at central location called hub**.
- In bus topology, all stations attach appropriate hardware interfacing known as a tap, directly to a linear transmission medium or a bus.
- **LAN** consists of a shared transmission medium and a set of hardware and software for interfacing devices to the medium.

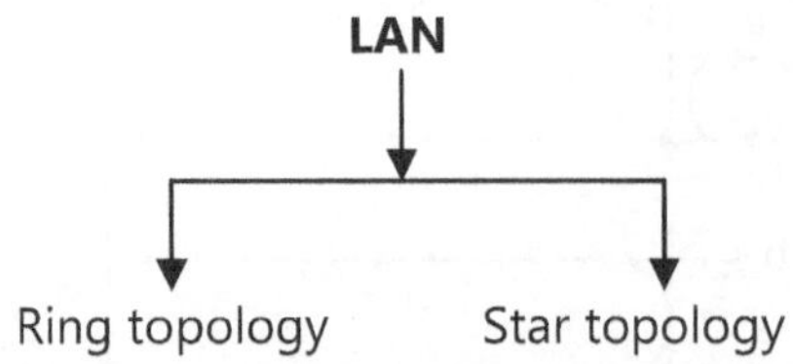

- In ring topology, consecutive nodes are connected by point-to-point links to form a closed ring.
- In star topology, each station/node is directly connected to a common, central node.
- Power budget is expressed as,

$$\bar{P}_{tr} = \bar{P}_{rec} + C_L + M_S$$

- **Long houl systems** are used to minimize the number of repeaters per unit distance.
- Line broadening is a **frequency chirp** associated with modulation induced charges in the carrier density.

Practice Questions

1. Explain lightwave system architecture.
2. Draw point-to-point link and explain.
3. Explain any one type of distribution networks.
4. Describe LAN and its types.
5. Draw loss-limited light wave system.
6. Explain dispersion limited light wave system.
7. Define long houl system.
8. Explain sources of power penalty.
9. Explain frequency chirping with diagram.
10. Give idea about reflection feedback noise.

***

Chapter 6

MULTI-CHANNEL SYSTEMS

6.1 INTRODUCTION

- Multiplexing divides the physical line or a medium into logical segments called *channels.*
- In multiplexing, different channels carry data simultaneously over same physical medium.
- Hardware equipment called multiplexer combine the inputs from different sources and load them on different channels of a medium.
- The combined data transverses over the medium simultaneously.
- At the receiving end, a demultiplexer separates the signals and sends them to their different destinations.
- Multiplexing is the transmission of information from one or more source to one or more destinations over the same transmission medium.
- The transmission medium may be metallic wire pair, a co-axial cable, a PCS mobile telephone, a terrestrial microwave radio system, a satellite microwave system or an fiber optic cable.
- There are several domains in which multiplexing can be accomplished, that includes space, phase, time, frequency and wavelength.
- Space Division Multiplexing (SDM) is a unsophisticated form of multiplexing that simply constitutes propagating signals from different sources on different cables that are contained within the same medium.
- The three most predominant methods of multiplexing are Time-Division Multiplexing (TDM), Frequency-Division Multiplexing (FDM) and Wavelength-Division Multiplexing (WDM).
- Subscriber multiplexing is depend on multiple microwave carriers for transmission of multiple signals.
- Code Division Multiplexing (CDM) is a networking technique in which multiple data signals are combined for simultaneous transmission over a common frequency bond.

6.2 MULTIPLEXING TECHNIQUES

- Whenever the bandwidth of a medium linking two devices if greater than the bandwidth needs of the devices, then the link can be shared.

Definitions:

- ***Multiplexing is the set of techniques that allows the simultaneous transmission of multiple signals across a single data link.***
- As data and telecommunication usage increases, so does the traffic also.
- We can install higher bandwidth links and use each to carry multiple signals.
- If the bandwidth of a link is greater than the bandwidth needs of the devices connected to it, the bandwidth is wasted.
- An efficient system maximizes the utilization of all resources, bandwidth is one of all the most precious resources we have in the data communication.
- In multiplexed system, n lines share the bandwidth of one link.
- Fig. 6.1 shows the basic format of a multiplexed system.

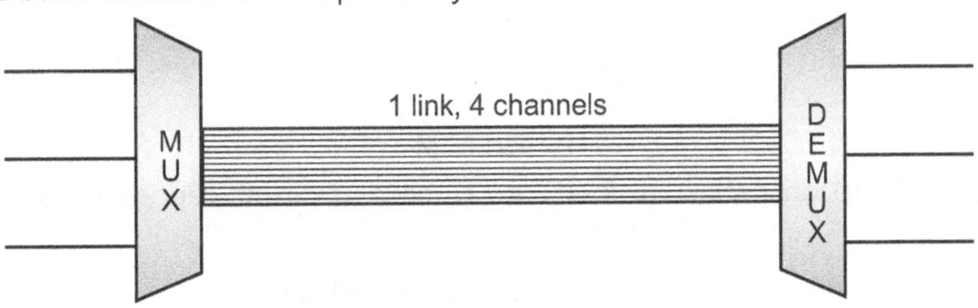

Fig. 6.1: Dividing a Link into Channels

- **Multiplexer** combines the four lines into one (many-to-one).
- At the receiving end, **Demultiplexer** separates the stream into four lines (one-to-many).
- **Need of Multiplexing:**
  - When many signals (channels) to be transmitted, then from transmitter side that sends simultaneously i.e. multiplexer coverts many into one so that at the receiving end also all inputs we get simultaneously sending many signals separately is expensive and requires more wires to send.

    For example, in cable TV, distributor sends many channels through single wire.
- **Types of Multiplexing are:**

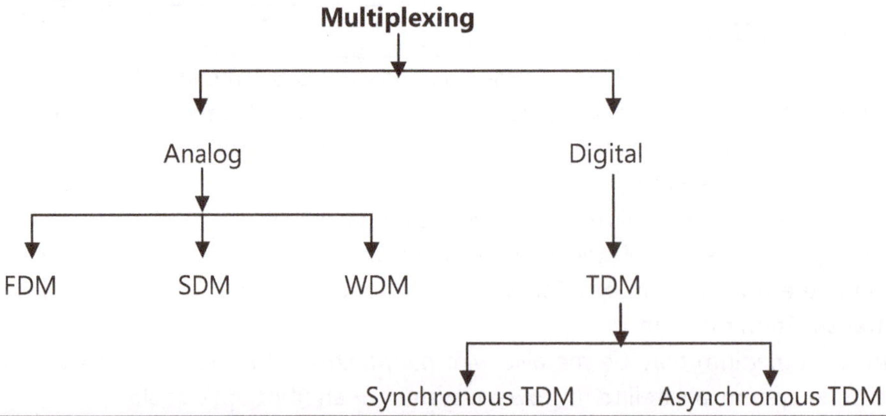

6.3 FREQUENCY-DIVISION MULTIPLEXING (FDM)

It is an analog technique.

Definition:

- *Frequency division multiplexing means total range of frequency is divided into number of frequency slots. Each slot of frequency is allotted to each channel.*
- **Principle:**

  Various channels of different frequencies combined, transmitted through single wire and separated at receiver with the help of demultiplexer.
- FDM can be applied when the bandwidth of a link is greater than the combined bandwidths of the signals to be transmitted.
- These modulated signals are then combined into a single composite signal that can be transported by the link.
- Carrier frequencies are separated by sufficient bandwidth to accommodate the modulated signal.
- These bandwidth ranges are the channels through which the various signals travels.
- Channels must be separated by guard bands to prevent signals from overlapping.
- Fig. 6.2 shows the conceptual view of FDM.

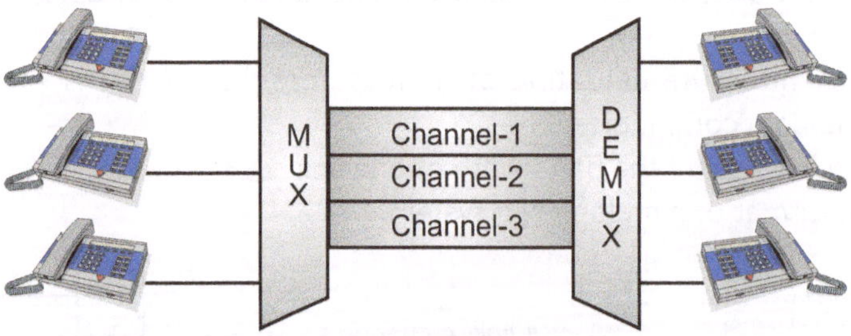

Fig. 6.2: FDM

From Fig. 6.2, the transmission path is divided into three parts, each representing a channel to carry one transmission.

Example, three narrow streets merge to form a three lane highway. Each of three streets corresponds to a lane of highway. Each car merging into the highway from one of the streets still has own lane and can travel without interfering with other cars.

6.3.1 FDM Transmitter (Multiplexing Process)

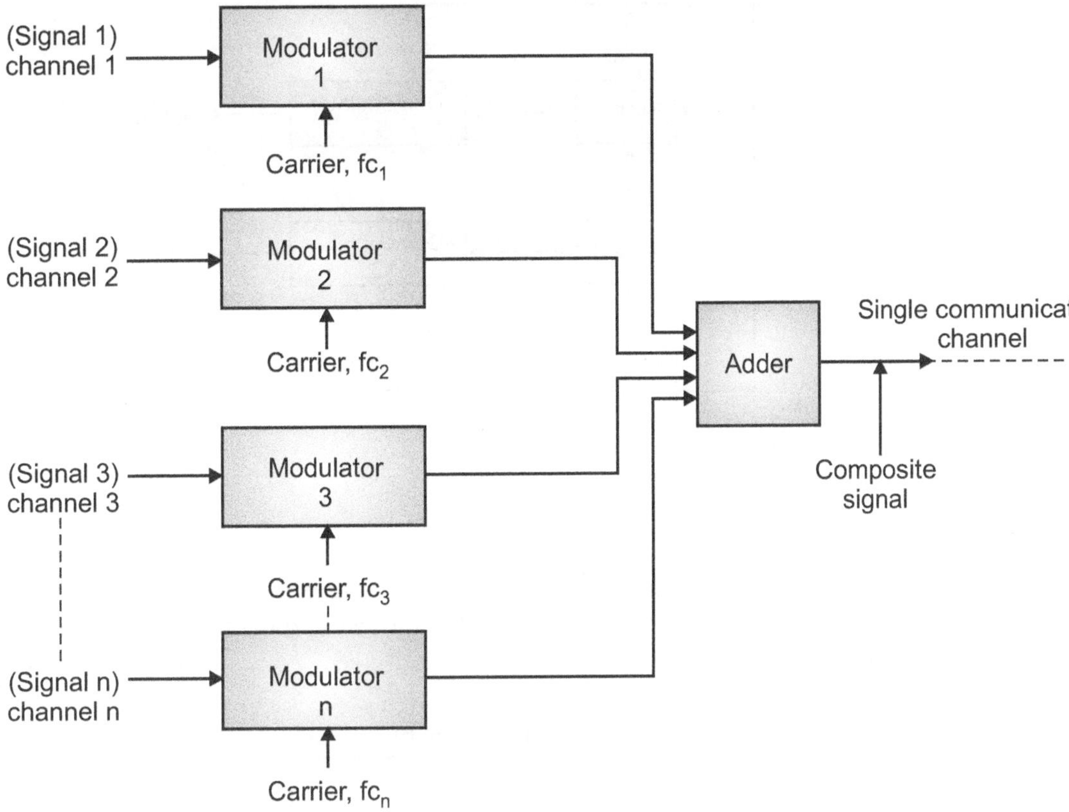

Fig. 6.3: FDM Transmitter

- As shown in Fig. 6.3, FDM is an analog process.
- Each signal is of similar frequency range.
- Each signal modulates a separate carrier and a modulator contains the sidebands of the corresponding signal.
- All the outputs from modulator added together in adder.
- Adder is a linear mixer which gives algebraic sum of all modulated signals.
- The resulting modulated signals are then combined into a signal that is sent over media link that has enough bandwidth to accommodate it.

6.3.2 FDM Receiver

- Demultiplexing process done at receiver.
- Demultiplexing means one into many.
- The composite signal is applied to different bandpass filters which separates individual channel.
- Each BPF has a center frequency corresponding to one of the carrier frequency.
- Each BPF pass only its channel and rejects all other channels.
- Each demodulator removes the corresponding carrier frequency and recovers the original signal back.
- The block diagram of FDM receiver is shown in Fig. 6.4.

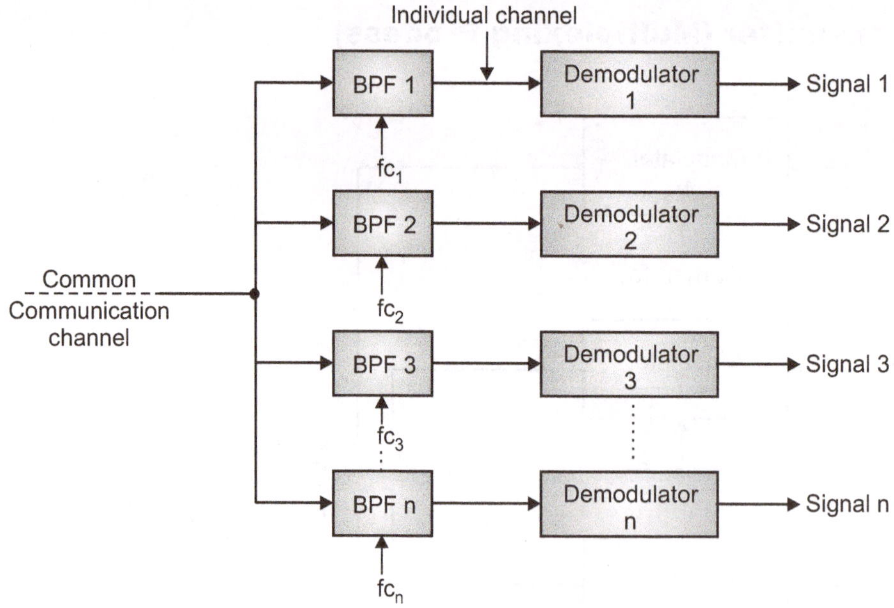

Fig. 6.4: FDM Receiver

Example 6.1:

Assume that a voice channel occupies a bandwidth of 4 kHz. We need to combine three voice channels into a link with a bandwidth of 12 kHz, from 20 to 32 kHz. Show the configuration using the frequency domain. Assume that there are no guard bands.

Solution:

We modulate each of three voice channels to a different bandwidth, as shown in Fig. 6.5.

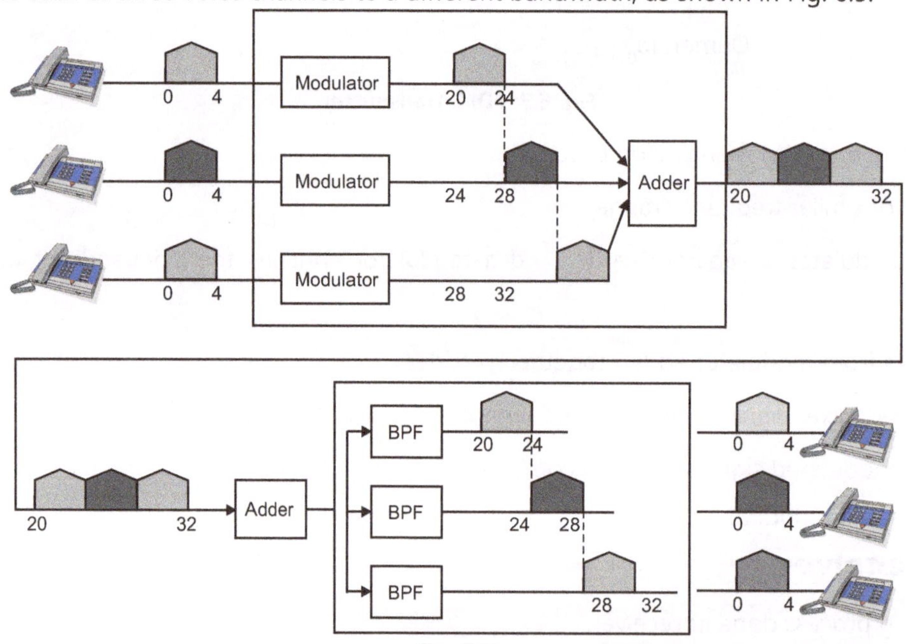

Fig. 6.5

- We use 20 to 24 kHz bandwidth for 1st, 24 to 28 kHz for 2nd and 28 to 32 kHz for 3rd channel.
- Then we combine these shown in Fig. 6.5.
- At the receiver, each channel receives the entire signal using a filter to separate out its own signal.

6.3.3 Advantages and Disadvantages and Applications of FDM

- **Advantages of FDM:**
  1. FDM is simpler and easy demodulation.
  2. Used for analog signals.
  3. Does not need synchronization between its transmitter and receiver.
  4. Due to slow narrow band fading only one channel gets affected.

- **Disadvantage of FDM:**
  1. Used only when a few low speed channels are desired.
  2. The communication channel must have large bandwidth.
  3. Intermodulation distortion takes place.
  4. FDM suffers problem of cross-talk.
- **Applications of FDM:**
  1. FDM is used to public telephones and in cable TV systems.
  2. AM and FM radio broadcasting.
  3. TV broad-casting.
  4. First generation of cellular phones.

Example 6.2:

Five channels, each with a 100 kHz bandwidth, are to be multiplexed together. What is minimum bandwidth of the link if there is a need for a guard band of 10 kHz between the channels to prevent interference?

Solution:

For five channels, 4 guard bands are required.

$$\therefore \quad \text{Required bandwidth} = 5 \times 100 + 4 \times 10 = 540 \text{ kHz}$$

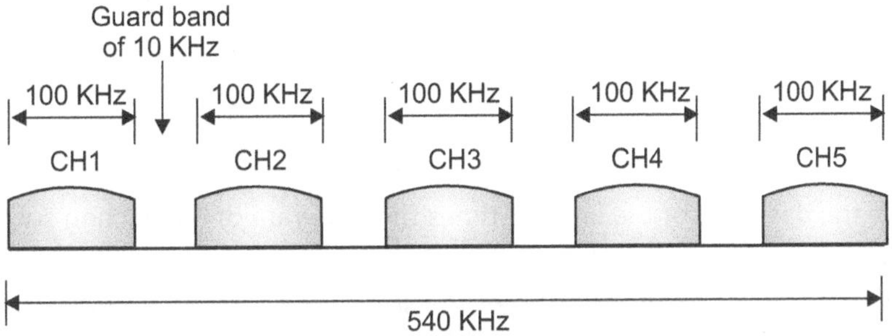

Fig. 6.6

6.4 TIME-DIVISION MULTIPLEXING (TDM)

- *TDM is digital multiplexing technique to combine data.*
- TDM is a process that allows several connections to share the high bandwidth of a link.
- Instead of sharing a portion of the bandwidths as in FDM, time is shared in TDM.

Principle:

Various channels having different time slots combined in a single wire and at the receiver they are separated with the help of demultiplexer.

- Fig. 6.7 shows conceptual view of TDM.

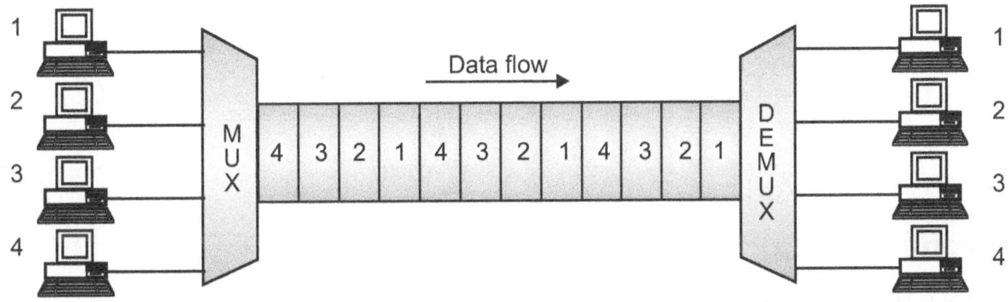

Fig. 6.7: TDM

- Same link is used as in FDM, but the link is shown sectioned by time rather than by frequency.
- In Fig. 6.9 portions of signals 1, 2, 3 and 4 occupy the link sequentially.
- We divide the transmission time into a number of time slices.
- We then allocate each time slice to different source nodes, each of which wants to send data.

- **Time Slots and Frames:**
  - The data flow of each connection is divided into units and the link combines one unit of each connection to make a frame.
  - The size of the unit can be 1-bit or several bits.

    Fig. 6.8 shows an example.
  - In TDM, the data rate of the link that carries data from n connections must be n times the data rate of a connection to guarantee the flow of the data.
  - Thus, the duration of a unit in a connection in n times the duration of a time slot in a frame.

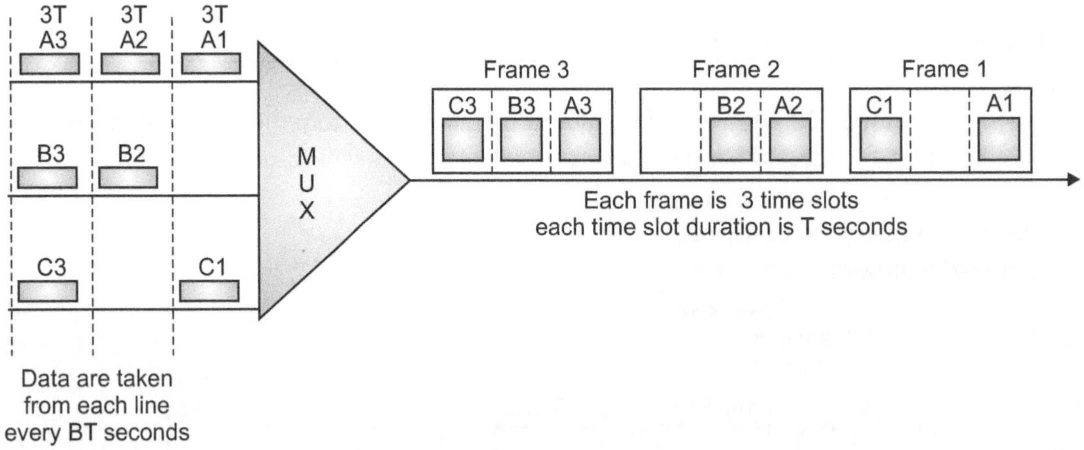

Fig. 6.8: TDM

- As shown in Fig. 6.8, the data rate of the link is 3 times the data rate of connection such that the duration of a unit on a connection is 3 times that of time slot.
- In Fig. 6.8, the data prior to multiplexing as 3 times the size of the data after multiplexing.
- This is just to convey the idea that each unit is 3 times longer duration before multiplexing than after.
- Time slots are grouped into frames.
- Here frame consists of one complete cycle of time slots with one slot decided to each sending device.
- In this system input lines are n, each frame has n slots, with each slot allocated to carrying data from a specific input line.

In such case, each node may want different amounts of data to be transmitted, because the time slice allocated to each source would be different. This requires a large number of buffer at each node to hold data when some other source node is transmitting. This method can be unsuitable to a few source, nodes at it may involve long waits if one of the source node **graphs** the medium for longer time.

For this reason or to overcome this problem the time span is divided into equal time slices.

So the TDM can be implemented in two ways.

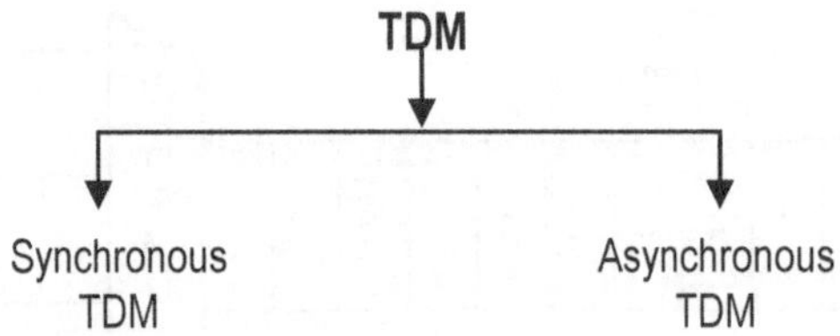

6.4.1 Synchronous TDM

The synchronous TDM also referred to as TDM.

The meaning of synchronous is different in other areas of telecommunication. But here it means that **the multiplexer allocates exactly the same time slots to each device at all times, whether or not a device has anything to transmit.**

For example, time slot A, is assigned to device A alone and not to used for any other device.

Fig. 6.9 shows synchronous TDM.

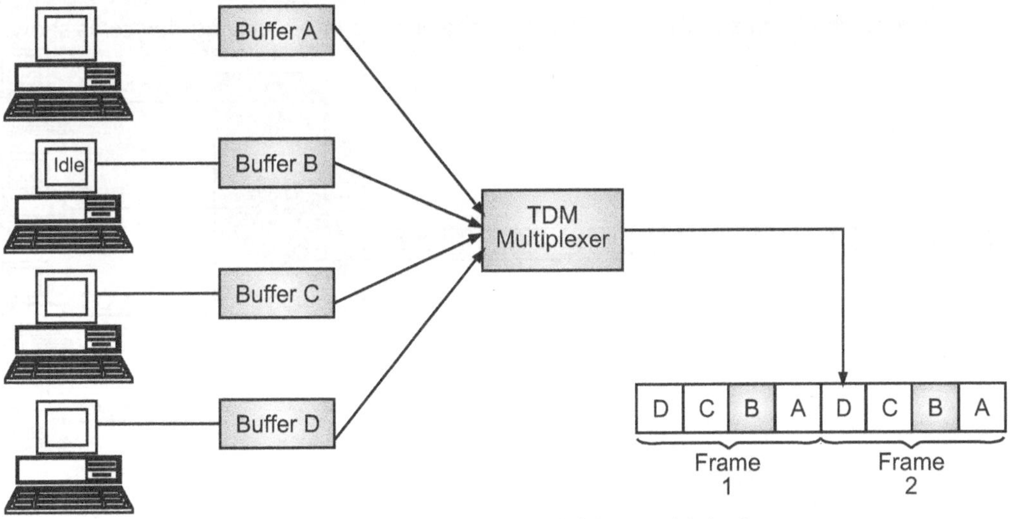

Fig. 6.9: Synchronous Time Division Multiplexing

- A small buffer memory is associated with each source node.
- At any time, not all the nodes may want to send some data.
- But the timing device in the multiplexer allocates some time for each node to transmit the data from its buffer as shown in figure.
- These time slots are grouped into frames. A frame consists of one complete cycle of time slots.
  For example, for n inputs, each frame has at least n slot, with each slot allocated to carrying data from a specific input line.
- Each device has one time slot per frame.
- After transmitting the data, the cycle repeats i.e. A-B-C-D-A-B-C-D etc.
- Then next turn comes, if a node wants to transmit any data.
- If there is no data to be transmitted, the buffer will be empty, but still the turn of the node will come shown in Fig. 6.11 (as node B).
- Here no addressing scheme to identify the source of the data but the position of the data in the composite message determine the source address.
- **Interleaving:**
  - o Synchronous TDM can be composed to a very fast rotating switch.
  - o As the switch opens in front of a device, that the device has the opportunity to send a specified amount of (bits) of data onto the path.
  - o The switch moves from device to device at a constant rate and in a fixed order. This process is called **interleaving**.

Interleaving can be done by bit, byte or any other data unit. The care must be taken that the interleaved units will always be of the same size.

Fig. 6.10 shows interleaving and frame building. Each device is sending different message. The multiplexer interleaves the different messages and forms them into frames before putting them onto the link.

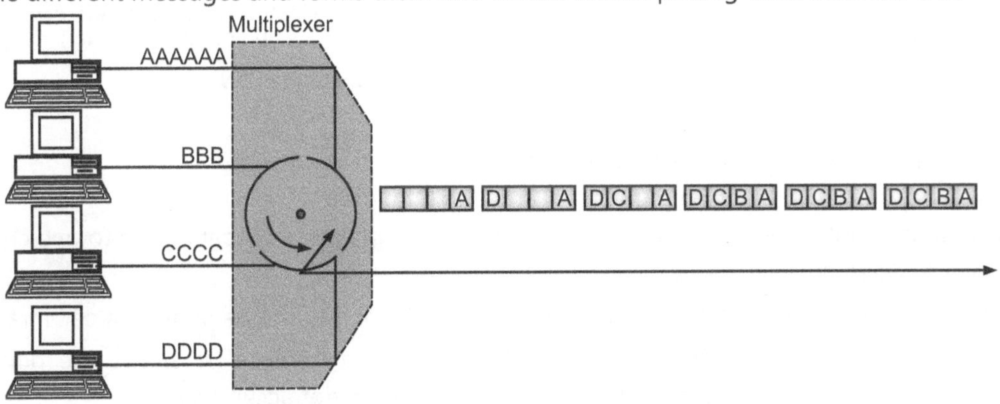

Fig. 6.10: Synchronous TDM, Multiplexing Process

At the receiver, the demultiplexer decomposes each frame by extracting each character in turn.

As a character is removed from a frame, it is passed to the appropriate receiving device as shown in Fig. 6.11.

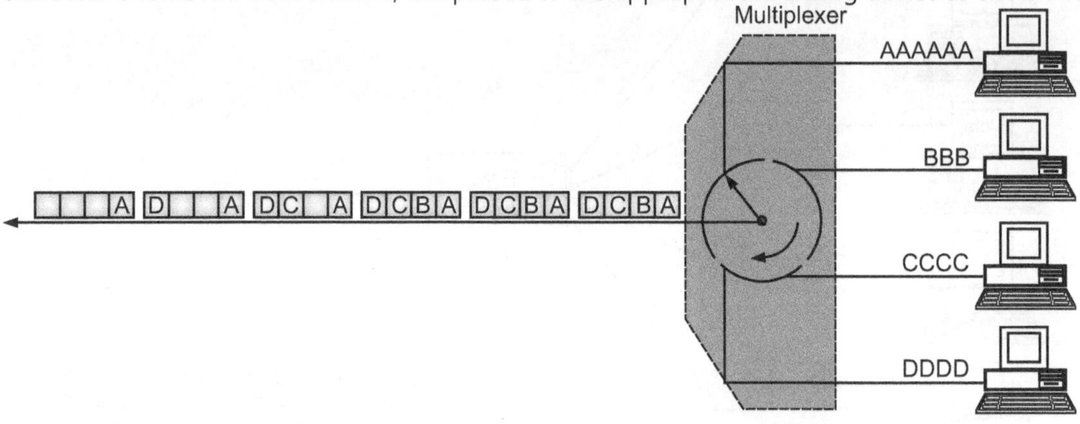

Fig. 6.11: Synchronous TDM, Demultiplexing Process

6.4.2 Asynchronous TDM

Here asynchronous means flexible or not fixed.

Asynchronous TDM is also known as **Statistical TDM** (STDM).

- This technique is more intelligent.
- It monitors which computer or terminal sending data more frequently and in greater quantity and allocates the time slices to those nodes.
- Inactive terminals/computers get the time slice less, while completely idle computer/terminal may not get any time slice at all.
- Only statistical data is kept about the activity of various source nodes and hence the name.

Fig. 6.12 shows the asynchronous TDM where the node B does not send any data at all while node C is more active, hence is **polled** more frequently.

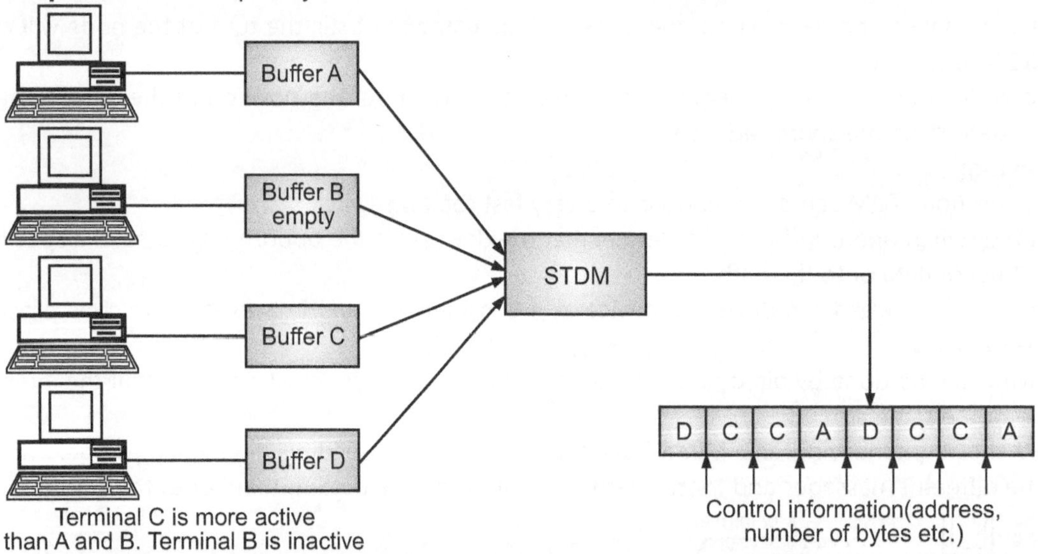

Fig. 6.12: Statistical Time Division Multiplexing (STDM)

6.4.3 Synchronizing

- Implementation of TDM is not easy as that of FDM.
- TDM needs synchronization between multiplexer and demultiplexer.
- If synchronization is not there between multiplexer and demultiplexer, a bit belonging to one channel may be received by the wrong channel.
- For this reason, one or more synchronization bits are usually added to the beginning of each frame.
- These bits are called **framing bits,** follow a pattern, frame to frame, that allows the demultiplexer to synchronize with the incoming stream so that it can separate time slot accurately.

- In most of the cases, this synchronization information consists of 1-bit per frame, changing between 0 and 1 as shown in Fig. 6.13.

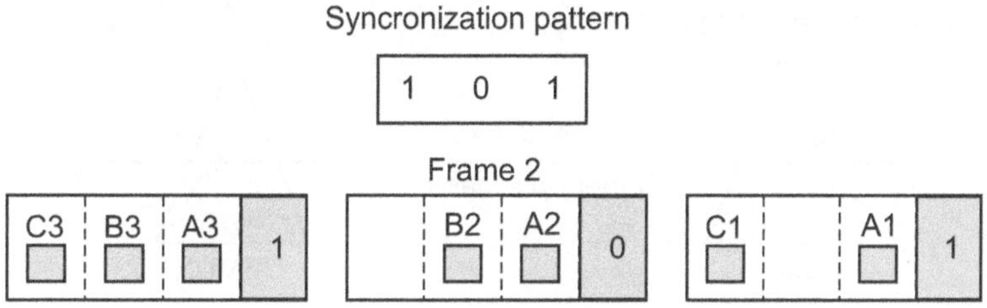

Fig. 6.13: Framing Bits

6.4.4 Advantages, Disadvantages and Applications of TDM

- **Advantages of TDM:**
    1. TDM systems are more flexible than FDM.
    2. Digital systems so often noise.
    3. No intermodulation distortion.
    4. Problem of cross-talk is not severe.
- **Disadvantage of TDM:**
    1. Complex to implement.
    2. Synchronization required.
- **Application of TDM:**
    1. In ISDN telephone lines.
    2. In wireline telephone systems.
    3. Some cellular telephone system.
    4. In circuit switched networks such as Public Switched Network (PSTN).

6.5 WAVE-DIVISION MULTIPLEXING (WDM)

- *WDM is an analog multiplexing technique to combine optical signals.*

Principle:

Very narrow bands of light from different sources are combined to make a wider band of light and at the receiver, the signals are separated by the demultiplexer.

- WDM is designed to use the high data rate capability of fiber optic cable.
- The optical fiber data rate is higher than the data rate of metallic transmission cable.
- Using a fiber optic cable for one single line wastes the available bandwidth.
- Multiplexing allows us to connect several lines into one.
- WDM is conceptually the same as FDM, except that the multiplexing and demultiplexing involve the optical signals transmitted through fiber optic cable.
- Fig. 6.14 shows the conceptual view multiplexer and demultiplexer.

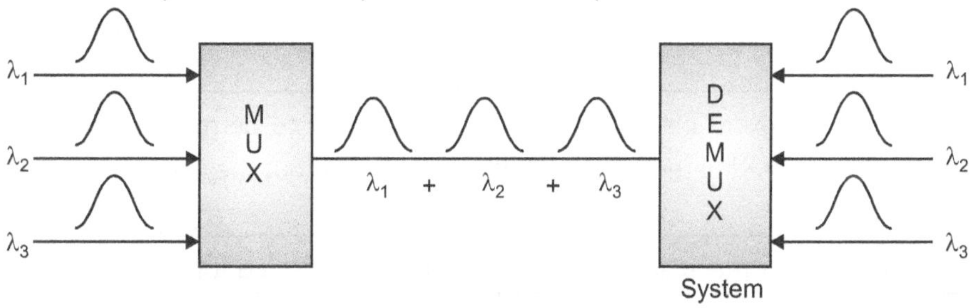

Fig. 6.14: WDM System

- Very narrow bands of light of differential wavelengths are combined to make a wide band of light.
- All wavelength travels through signal cable.
- At the receiver, the signals are separated by the demultiplexer.
- Combining and splitting of light sources are easily handled by a prism.
- We studied in Physics, that a prism bends a beam of light based on the angle of incidence and the frequency.

- Using this technique, a multiplexer can be made to combine several input beams of light, each containing narrow band of frequencies into one output beam of wider band of frequencies.
- A demultiplexer does the reverse process.

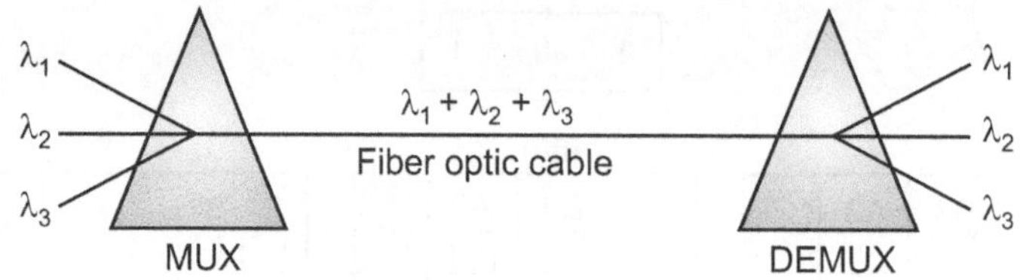

Fig. 6.15: Prisms with WDM

- **Advantages of WDM:**
  1. In low bandwidth large number of channels accommodated.
  2. Time sharing not required.
  3. High rate.
  4. High speed.
- **Disadvantage of WDM:**
  Only possible with fiber optic cable which is expensive.
- **Applications of WDM:**
  - WDM used in SONET (synchronous optical network) where SONET is a high-data-rate network for fiber optic networks.

6.6 CODE DIVISION MULTIPLEXING (CDM)

Code Division Multiplexing (CDM) is a networking technique in which multiple data signals are combined for simultaneous transmission over a common frequency band.

In CDM separation is achieved by assigning each user channel its own code. Guard spaces are realized using codes with necessary distance in code spaces, orthogonal codes. When CDM is used to allow multiplie users to share a single communication channel, the technology called CDMA.

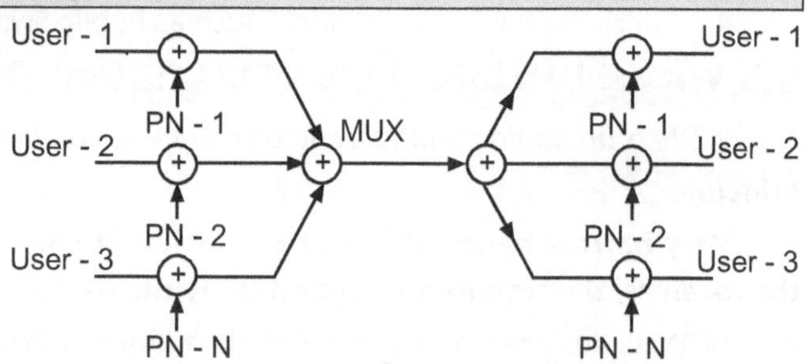

Fig. 6.16: Code Division Multiplexing for N-users

Advantage:

Good protection against unauthorized reception.

6.6.1 Comparison of FDM and TDM

Sr. No.	FDM	TDM
1.	The signals to be multiplexed are added in time domain. But they occupy different slots in the frequency domain.	The signals to be multiplexed can occupy the entire bandwidth but they are isolated in the time domain.
2.	Mostly FDM is preferred for the analog signals.	TDM is preferred for digital signals.
3.	In FDM, synchronization is not required.	In TDM, synchronization is required.
4.	FDM suffers from cross-talk like problems due to imperfect band pass filter.	In TDM cross-talk like problem not occurs.
5.	Due to wideband fading in the transmission medium, all the FDM channels are affected.	Due to fading, only a few TDM channels will be affected.

6.6.2 Subcarrier Multiplexing (SCM)

- The basic concept behind subcarrier multiplexing is taken from microwave technology, which employs multiple microwave carriers for transmission of multiple signals over coaxial cables or free space.

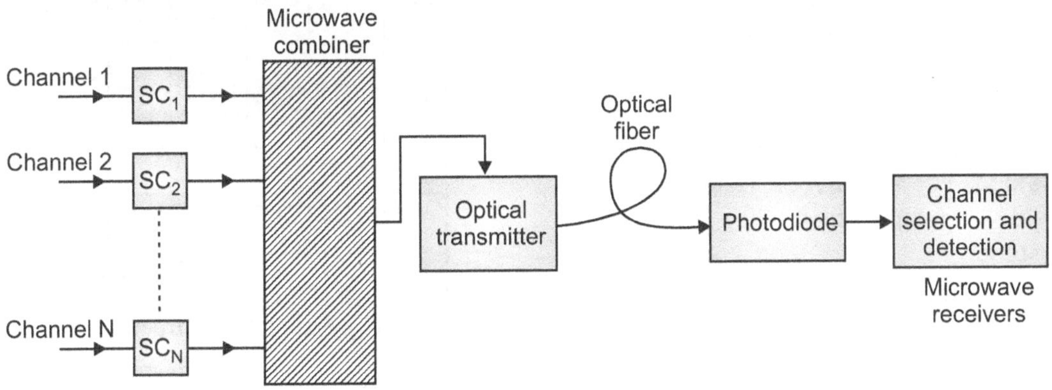

Fig. 6.17: Subcarrier Multiplexing

- The total bandwidth is limited to below 1 GHz when coaxial cables are used to transmit a multichannel microwave signal. However, if the microwave signal is transmitted optically by using optical fiber, the signal bandwidth can easily exceed 10 GHz for a single optical carrier.
- Such a scheme is referred to as SCM, since multiplexing is done by using microwave subcarriers rather than optical carrier.
- Fig. 6.17 shows a SCM tightwave system designed with a single optical carrier. The main advantage of SCM is the **flexibility** and **upgradability**.

6.6.3 Code Division Multiple Access (CDMA)

- Coding Division Multiplexing (CDM) is a networking technique in which multiple data signals are combined for simultaneous transmission over a common frequency band.

Code Division Multiplexing (CDM):

- In code division multiplexing, each channel is coded in such a way that, its spectrum spreads over a much wider region that occupied by the original signal.
- The new components needed for CDM systems are encoders and decoders located at the transmitter and receiver ends respectively.
- The encoder spreads the signal spectrum over a much wider region that the minimum bandwidth necessary for transmission.
- The decoder uses the same code for compressing the signal spectrum and recovering the data. The spectrum-spreading code is called signature sequence.
    1. In CDM separation is achieved by assigning each user channel its own code. Guard spaces are realized using codes with necessary distance in code spaces, orthogonal codes. When CDM is used to allow multiplied users to share a single communication channel, the technology called CDMA.
    2. Good protection against unauthorized reception is the main advantage of CDM.
    3. A receiver must be precisely synchronized with the transmitter to apply the decoding correctly. All the signals received by the receiver should have equal strength, otherwise some signals drain others.

Advantage:

1. Good protection against unauthorized reception.

By combining TDMA and FDMA techniques, a hybrid technique formed called CDMA. Frequency hoping can be used so that the frequency bands assigned to the users are recorded in an essentially random manner.

Fig. 6.18 (b) shows frequency time structure of CDMA technique.

From structure, during time slot 1, user 1 occupies frequency band 1, user 2 occupies frequency band 2 and user 3 occupies frequency band 3. During second time slot, user 1 hops to frequency band 1, user 3 hops to band 2 and so on.

- Example, suppose in a room, everyone is talking at once, then nothing can be understood. But if everyone is talking in different languages, then they can be understood.
- Like this, CDMA has no channels but instead, every call is encoded by adding certain code (called psudo-noise code) that makes it different from other calls in frequency spectrum.

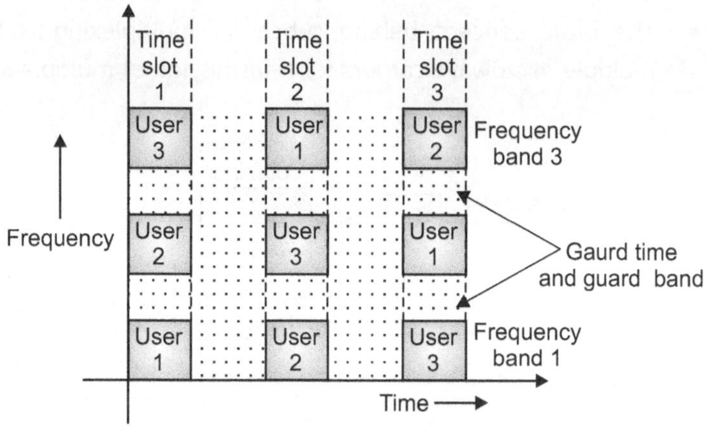

Fig. 6.18: CDMA Structure

- Similar to FDMA and TDMA multiple access can be organised around code division multiplexing CDMA.
- In CDMA, individual bits of a message are encoded in terms of unique digital codes.
- Many stations, each having its unique code, transmit their signals simultaneously.
- The signals are discriminated by means of a cross correlator, using the code of the desired transmission as the reference code.
- Since each transmitted symbol is encoded as a sequence of higher rate symbols, the scheme demands a much larger bandwidth for transmission and the technique is called code division multiple access, and hence is referred to as spread spectrum technique.
- The number of higher rate coding symbols per bit usually called chips, is an important parameter in CDMA, and is referred to as processing gain or bandwidth spreading factor.
- The spectrum spreading property significantly reduces interference from other transmission.
- The codes used for CDMA are specifically selected to have good cross-correlation properties so that interference from other unwanted signals simultaneously may be minimised.
- A family of pseudo random (PR) codes, known as GOLD codes have good cross correlation properties.
- CDMA may be used in two ways, one way is that every transmitting station uses its own PR sequence encoding data and a receiver station uses a separator correlator for each station it wishes to receive.
- The other way is for the transmitting station to choose the PR sequence of the destination station to encode its message.
- In this case each receiver station has only one correlator using its own PR code as the reference sequence.
- One of the applications, for which CDMA is very suitable is to transmit low bit rate data from a central station to any of a large number of receive only small stations.
- The capability of CDMA to selectively address the remote stations by use of different codes comes in very handy here.
- In India **NICNET** and **RABMN use a CDMA access technique**.

Advantages of CDMA:

1. Any external synchronization not required.
2. Noise reduction capability is better.

Important Points

- **Multiplexing** is the set of techniques that allows the simultaneous transmission of multiple signals across a single data link.
- **Multiplexer** means many into one.

- **Demultiplexer** means one into many.

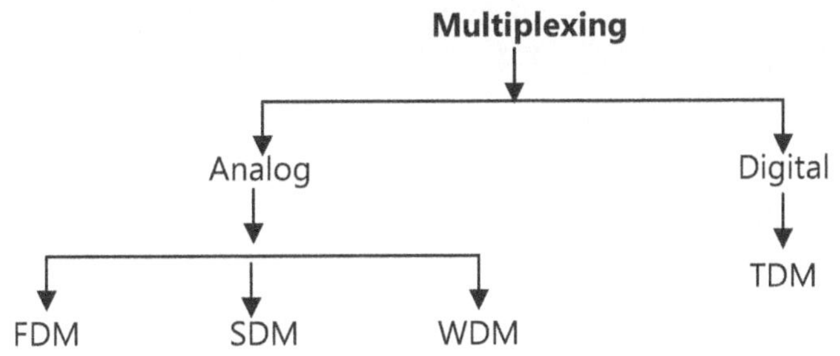

- **FDM** means total range of frequency is divided into number of frequency slots.
- **WDM** is an analog multiplexing technique to combine optical signals.
- **TDM** is a digital multiplexing technique to combine data.
- In a **TDM,** the data rate of the link is n times faster and the unit duration is n times shorter.
- The satellite transponder can be accessed by several earth stations from different locations on earth simultaneously. This is called **Multiple Access.**

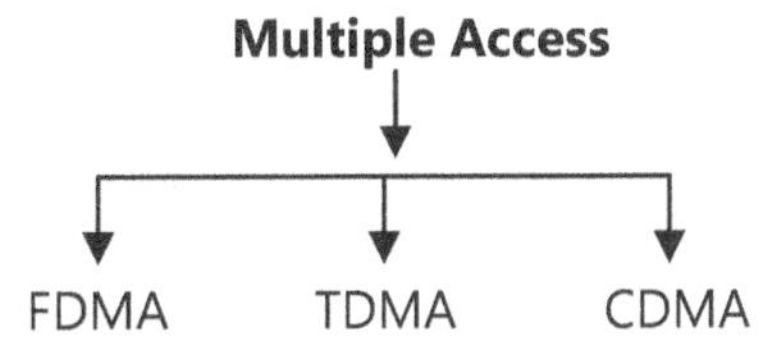

- The satellite transponder can be accessed by several earth stations from different locations on earth simultaneously. This is called multiple access.
- In TDMA technique, each user is given a particular time slot regularly after fixed interval. During this slot, the user have the complete availability of satellite transponder.
- By combining TDMA and FDMA techniques, a hybrid technique formed called CDMA.
- The SDMA uses the directional antenna for splitting the coverage.
- The OFDM is a digital modulation technique that can support high-speed video communication along with audio.
- Orthogonality between two signals means that the two coexisting signals are independent of each other in a specified time interval and do not interact with each other.
- Orthogonality is a property that allows multiple information signals to be transmitted perfectly over common channel and detected without interference.

Practice Questions

1. What is the need of multiplexing?
2. State the principle of FDM and explain the concept.
3. State the principle of WDM and explain any one.
4. List the types of TDM and explain.
5. Explain synchronous TDM.
6. Explain asynchronous TDM.
7. Compare FDM and TDM.
8. Explain the concept of CDMA.
9. Explain subcarrier multiplexing.

***